Concepts in Competitive Mathematics

Second Edition

Zachary M. Boazman

Edited by Dr. Richard Newcomb

Cistercian Preparatory School

Irving, Texas

Copyright 2014
Zachary Boazman
Cistercian Preparatory School

For the Cistercian Class of 2010

Contents

1 Number Theory   **7**
    1.1 Divisibility Results 7
    1.2 Testing for Primality 9
    1.3 Prime Factorization and its Utility 10
    1.4 The GCD 12
    1.5 The LCM 13

2 Combinatorics   **15**
    2.1 Basic Combinatorics Principles 18
    2.2 Permutations 19
    2.3 Combinations 20
       2.3.1 Combinations with Repetition: "Dashes and Dividers"   22
       2.3.2 Other Interpretations of $\binom{n}{k}$ 25
    2.4 Recurrence Relations 26
    2.5 Pigeonhole Principles 27
    2.6 How to Avoid Over Counting 28
    2.7 Multifaceted Examples in Combinatorics 30

3 Probability and Statistics   **37**
    3.1 The Basics of Probability 37
    3.2 Probability of Multiple Events 40
    3.3 Basic and Complex Examples in Probability 41
    3.4 Statistics 43
       3.4.1 Data Distribution 43
       3.4.2 Three Kinds of Means 44
       3.4.3 Examples in Statistics 46

4 Sequences and Series   **49**
    4.1 Arithmetic Sequences and Series 50
    4.2 Geometric Sequences and Series 52
    4.3 Special Partial Sums and Series 53
    4.4 The Fibonacci Sequence 54

5 Algebra — 57
- 5.1 Polynomials 57
  - 5.1.1 Quadratic Functions 57
  - 5.1.2 Cubic Functions 59
  - 5.1.3 Polynomial Properties 60
- 5.2 Exponents and Logarithms 62
  - 5.2.1 Exponents 62
  - 5.2.2 Logarithms 63
  - 5.2.3 Exponent and Logarithm Examples 63

6 Geometry — 65
- 6.1 Circles 66
- 6.2 Quadrilaterals 68
- 6.3 Triangles 70
- 6.4 Angles 74
- 6.5 Three Dimensional Solids 74
- 6.6 Geometry Examples 76

7 Trigonometry — 81
- 7.1 Definitions 81
- 7.2 Basic Values and Identities 82
- 7.3 Law of Sines and Law of Cosines 83
- 7.4 Examples 84

8 The Coordinate Plane — 87
- 8.1 Lines 87
- 8.2 Parabolas 88
- 8.3 Polygon Tricks 88
- 8.4 Miscellaneous 90

9 Problem Solving Strategies — 91
- 9.1 Number Theory Tips 91
- 9.2 Combinatorics Tips 92
- 9.3 Probability and Statistics Tips 92
- 9.4 Sequences and Series Tips 93
- 9.5 Algebra Tips 94
- 9.6 Geometry Tips 94
- 9.7 Trigonometry Tips 95

10 Examples Cited — 97

Index — 99

Chapter 1

Number Theory

All variables in this section are integers.

1.1 Divisibility Results

NOTATION 1.1   If m divides evenly into n we denote this fact as '$m \mid n$'.

FORMULA 1.1   **Basic Divisibility Properties.**

For all integers, we have the following basic divisibility properties:

- $m \mid 0$ always.
- $m \mid m$.
- If $m \mid n$ and $n \mid k$, then $m \mid k$.
- If $m \mid n$ and $m \mid k$, then $m \mid \alpha n + \beta k$.
- If $m \mid n$ and $m \mid n \pm k$, then $m \mid k$.
- If $m \mid n$ and $n \mid m$, then $|n| = |m|$.
- If $m \mid n$ and $n \neq 0$, then $\frac{n}{m} \mid n$.
- Given $k \neq 0$, $m \mid n$ if and only if $mk \mid nk$.

FORMULA 1.2   **Divisibility Rules for Selected Integers.**

- $2 \mid m$ exactly when the last digit of m is even (i.e. divisible by 2).

- $3 \mid m$ exactly when 3 divides the sum of the digits of m.

  EXAMPLE 1.1 The integer 27435 is divisible by 3 because $2+7+4+3+5 = 21$ and $3 \mid 21$.

- $4 \mid m$ exactly when four divides the last two digits of m. When testing the divisibility of two digit numbers, subtracting a multiple of 20 does not affect divisibility.

  EXAMPLE 1.2 The integer 3492 is divisible by 4 because $4 \mid 92$ and $4 \mid 92$ because $92 - (4 \times 20) = 12$ and $4 \mid 12$.

- $5 \mid m$ exactly when m ends in a zero or a five.

- $6 \mid m$ exactly when m is divisible by both 2 and 3.

  EXAMPLE 1.3 The integer 39258 is divisible by six because it ends in an even digit and $3+9+2+5+8 = 27$ and $3 \mid 27$.

- $7 \mid m$ exactly when the number resulting when twice the last digit is subtracted from the rest of the number is divisible by seven.

  EXAMPLE 1.4 The integer 931 is divisible by 7 because $93 - 2(1) = 91$, and then $9 - 2(1) = 7$.

- $8 \mid m$ exactly when eight divides the last three digits of m. When testing the divisibility of three digit numbers, subtracting a multiple of 40 or 200 does not affect divisibility.

  EXAMPLE 1.5 The integer 76536 is divisible by 8 because $536 - (2 \times 200) = 136$ and $136 - (3 \times 40) = 16$ and $8 \mid 16$.

- $9 \mid m$ exactly when nine divides the sum of the digits of m.

  EXAMPLE 1.6 The integer 41382 is divisible by 9 because $4+1+3+8+2 = 18$ and $9 \mid 18$.

- $11 \mid m$ exactly when the alternating sum and difference of the digits of m is divisible by 11.

  EXAMPLE 1.7 The integer 187 is divisible by 11 because $-1+8-7 = 0$ and $11 \mid 0$.

1.2. TESTING FOR PRIMALITY

EXAMPLE 1.8   The integer 8362948 is divisible by 11 because $8 - 3 + 6 - 2 + 9 - 4 + 8 = 22$ and $11 \mid 22$.

- $13 \mid m$ exactly when the number resulting when 4 times the last digit is added to the rest of the number is divisible by 13.

EXAMPLE 1.9   The integer 637 is divisible by 13 because $63 + (7 \times 4) = 91$ and $9 + (1 \times 4) = 13$ and $13 \mid 13$. Note, therefore, 91 is not prime!

- $17 \mid m$ exactly when the number resulting when 5 times the last digit is subtracted from the rest of the number is divisible by 17.

EXAMPLE 1.10   The integer 867 is divisible by 17 because $86 - (7 \times 5) = 51$ and $5 - (1 \times 5) = 0$ and $17 \mid 0$.

- $19 \mid m$ exactly when the number resulting when twice the last digit is added to the rest of the number is divisible by 19.

EXAMPLE 1.11   The integer 893 is divisible by 19 because $89 + (2 \times 3) = 95$ and $9 + (2 \times 5) = 19$ and $19 \mid 19$.

1.2 Testing for Primality

For a positive integer n to be prime, it is enough that no prime less than or equal to $\sqrt{n}$ divides n.

EXAMPLE 1.12   **Is 439 prime?**
Solution: Note: $400 < 439 < 441$. Therefore, $20 < \sqrt{439} < 21$. Hence, to test for primality, we must only test all primes up to 20.

- $2 \nmid 439$ because the last digit isn't even.
- $3 \nmid 439$ because $4 + 3 + 9 = 16$ and $3 \nmid 16$.
- $5 \nmid 439$ because the last digit isn't a 0 or 5.
- $7 \nmid 439$ because $43 - 18 = 25$ and $7 \nmid 25$.
- $11 \nmid 439$ because $4 - 3 + 9 = 10$ and $11 \nmid 10$.

- $13 \nmid 439$ because $43+(4\times 9) = 79$ and $7+(4\times 9) = 43$ and $13 \nmid 43$.
- $17 \nmid 439$ because $43 - (5 \times 9) = -2$ and $17 \nmid -2$.
- $19 \nmid 439$ because $43 + (2 \times 9) = 61$ and $6 + (2 \times 1) = 8$ and $19 \nmid 8$.

Thus, we have concluded that because 439 has failed all divisibility tests on primes less than its square root, 439 is prime.

1.3 Prime Factorization and its Utility

THEOREM 1.1   **Fundamental Theorem of Arithmetic.**

Any integer n greater than 1 has a unique representation (up to a permutation) as a product of primes.

DEFINITION 1.1   **Prime Factorization.**

The prime factorization of n is the unique representation of n as a product of primes, i.e. $n = p^a q^b \ldots u^f$.

EXAMPLE 1.13   **What is the Prime Factorization of 488796?**
Solution: Find the prime factorization by dividing powers of primes. Tip: test primes with easier divisibility tests first.

- Begin by noticing that $4 \mid 96$. Therefore, divide out the factor of 4 or 2^2 to get 122199.

- Next, note that $1+2+2+1+9+9 = 24$ and $3 \mid 24$. Therefore, divide out the factor of 3 to get 40733.

- Next, note that $4 - 0 + 7 - 3 + 3 = 11$ and $11 \mid 11$. Therefore, divide out the factor of 11 to get 3703.

- Next, note that $370 - 3 \times 2 = 364$ and $36 - 2 \times 4 = 28$ and $7 \mid 28$. Therefore divide out the factor of 7 to get 529.

- Finally, note that 529 is 23^2, so note that as the last prime factor.

- Collect all factors to express the prime factorization of 488796 as $2^2 \cdot 3 \cdot 7 \cdot 11 \cdot 23^2$.

1.3. PRIME FACTORIZATION AND ITS UTILITY

To do this quickly use inverted division.

```
 2 | 488796
 2 | 244398
 3 | 122199
11 |  40733
 7 |   3073
23 |    529
           23
```

and recollect all the factors.

FORMULA 1.3 Number of Factors of n.

If the prime factorization of n is $p^a q^b \ldots u^f$ then the number of factors of n is $(a+1)(b+1)\cdots(f+1)$.

This formula is actually a combinatoric application to number theory. Think of it this way: each factor has $a+1$ choices for how many p's to have in its prime factorization, $b+1$ choices for how many q's to have in its prime factorization, etc.[1] By the MPC (see below), we can multiply these choices together to get the total number of choices.

EXAMPLE 1.14 How many factors of 11088 are there?

The prime factorization of 11088 is $2^4 \cdot 3^2 \cdot 7 \cdot 11$. Thus, the number of factors is $(4+1) \cdot (2+1) \cdot (1+1) \cdot (1+1) = 5 \cdot 3 \cdot 2 \cdot 2 = 60$.

FORMULA 1.4 Sum of the Factors of n.

Let $\mathcal{S}(n)$ be the sum of the factors (positive) of n (this includes 1 and n). First note that if n and m are relatively prime (see below), then $\mathcal{S}(n \cdot m) = \mathcal{S}(n) \cdot \mathcal{S}(m)$. Second $\mathcal{S}(p^r) = 1 + p + p^2 + \ldots + p^r$. Combining these two facts allows one to quickly compute $\mathcal{S}(n)$.

EXAMPLE 1.15 What is the sum of all the factors of 200?

$\mathcal{S}(200) = \mathcal{S}(8 \cdot 25) = \mathcal{S}(2^3) \cdot \mathcal{S}(5^2) = (1+2+4+8)(1+5+25) = 15 \cdot 31 = 465$.

---

[1] It's always $a+1$ because there can be 1 or 2 or 3 or ... a p's in the prime factorization or no p's at all.

1.4 The GCD

DEFINITION 1.2   **Greatest Common Divisor (GCD).**

Let D_m and D_n be the set of all of the factors of m and n respectively. The largest element in the intersection $D_m \bigcap D_n$ is called the Greatest Common Divisor or GCD of m and n.

DEFINITION 1.3   **Relatively Prime.**

Two integers m and n are relatively prime if their Greatest Common Divisor is 1.

Practically speaking, this means that two numbers are relatively prime if they don't share any of the same prime factors.

DEFINITION 1.4   **The Totient of n.**

The totient of n is the number of positive integers less than or equal to n that are relatively prime to n. Notationally, this is called $\varphi(n)$.

EXAMPLE 1.16   **How many positive integers less than or equal to 28 are relatively prime to 28?**
Solution: The prime factorization of 12 is $2^2 \cdot 7$. The integers that don't have any of these factors are 1, 3, 5, 9, 11, 13, 15, 17, 19, 23, 25, and 27. That's 12 integers.

EXAMPLE 1.17   **How many positive integers less than or equal to 280 are relatively prime to 280?**
It would be imprudent to list said integers. A totient formula is needed, therefore.

FORMULA 1.5   **Euler's Totient (or Phi) Function**

Let $p^a q^b \ldots u^f$ be the prime factorization of n. The number of positive integers less than or equal to n that are relatively prime to n can be expressed as

$$\varphi(n) = \left[(p-1) \cdot (p^{a-1})\right] \times \left[(q-1) \cdot (q^{b-1})\right] \times \cdots \times \left[(u-1) \cdot (u^{f-1})\right].$$

Note that when the $\gcd(m,n) = 1$, the totient function is multiplicative, that is, $\varphi(nm) = \varphi(n) \cdot \varphi(m)$.

1.5. THE LCM

Solution: To solve the question posed above, we'll use the Euler's formula. The prime factorization of 280 is $2^3 \cdot 5 \cdot 7$. Applying the above formula, we get the number of relatively prime integers less than or equal to 280 to be

$$\left[(2-1) \cdot (2^{3-1})\right] \times \left[(5-1) \cdot (5^{1-1})\right] \times \left[(7-1) \cdot (7^{1-1})\right]$$
$$= 1 \times 4 \times 4 \times 1 \times 6 \times 1 = 96.$$

1.5 The LCM

DEFINITION 1.5   **Least Common Multiple (LCM).**

Let M_m and M_n be the set of all of the multiples of m and n respectively. The smallest positive element in the intersection $M_m \bigcap M_n$ is called the Least Common Multiple or LCM of m and n.

FORMULA 1.6   **The GCD-LCM Product.**

The product $m \cdot n$ is equal to the product of the GCD and the LCM of m and n. That is

$$mn = \text{lcm}(m,n) \cdot \gcd(m,n).$$

EXAMPLE 1.18   **The product of two positive integers is 216. Their LCM is 36. What is the greatest factor common to both integers?**

Solution: Let g be the GCD of the two integers. Then use the formula above.

$$36g = 216$$
$$g = 6$$

The GCD of the two integers, therefore, is 6.

Chapter 2

Combinatorics

Combinatorics is a branch of mathematics which studies finite or countably infinite discrete structures. Although it may seem like a straightforward problem to count objects one should consider:

1. Which "objects" are considered distinct?

2. How can the counting be done efficiently and quickly?

Some answers to these questions are given in the problems, formulas and examples below.

EXAMPLE 2.1   Suppose each student in one film class were asked to rank a set of four classic drama movies: *Citizen Kane*, *12 Angry Men*, *To Kill A Mockingbird*, and *Chinatown*, with no ties allowed. Then, suppose each student in another film class, none of the students being the same, were asked to rank a set of five classic comedy movies: *Blazing Saddles*, *Airplane*, *The Blues Brothers*, *The Pink Panther (1963)*, and *Breakfast at Tiffany's*, again with no ties allowed. How many different total rankings are possible? What is the smallest class size for the first class which would guarantee that at least two students gave the same ranking of the selected drama movies?

Solution: To answer the first question, it seems natural to divide the "total rankings" into its two constituent parts: the ranking of the dramas and the ranking of the comedies. Having found the number of rankings possible for each category, we can add them together using the **Simple Addition Principle (SAP)** of counting.

First the Dramas. When choosing the top ranked drama movie there are only four possibilities since no ties are allowed. After the top movie is chosen, there are only three choices left for the second ranked movie. For the third ranked there will be two choices; for the fourth ranked only one movie is left to "choose". How do we count all possible rankings? If we use the **Multiplication Principle of Counting (MPC)** described below we have

$$\# \text{ of drama rankings}$$
$$= \underline{4} \cdot \underline{3} \cdot \underline{2} \cdot \underline{1}$$
$$= 24.$$

So there are 24 possible rankings of the four drama movies. You can see why the MPC works in the tree diagram on the following page. Each rectangle contains a ranked movie, and each rectangle then has a number of lines to indicate the number of choices for the next ranked movie.

Similarly, for the comedies, we have five choices for the top movie, four for the second, three for the third, two for the fifth, and one for the least favorite movie. Again using the MPC, we have

$$\# \text{ of comedy rankings}$$
$$= \underline{5} \cdot \underline{4} \cdot \underline{3} \cdot \underline{2} \cdot \underline{1}$$
$$= 120.$$

So there are 120 possible rankings of the five comedy movies.

Therefore, using the SAP, the answer to the first question, the total number of rankings, is $24 + 120 = 144$ total rankings.

In combinatorics problems, the MPC is often executed through the use of "the hangman method". Dashes indicate a choice and numbers are used to indicate the number of choices. Using the MPC (see below), the numbers are multiplied to give the total number of choices.

The mathematical term for ranking is **permutation**. In a permutation the order of the elements being selected is crucial.

As to the second question, imagine as many students as possible having different rankings. That must be 24 students. So a class of 25 would be the smallest class size which would have to have at least two students with the same rankings. This reasoning is an example of the **Pigeonhole Principle** given below.

Tree Diagram for Dramas

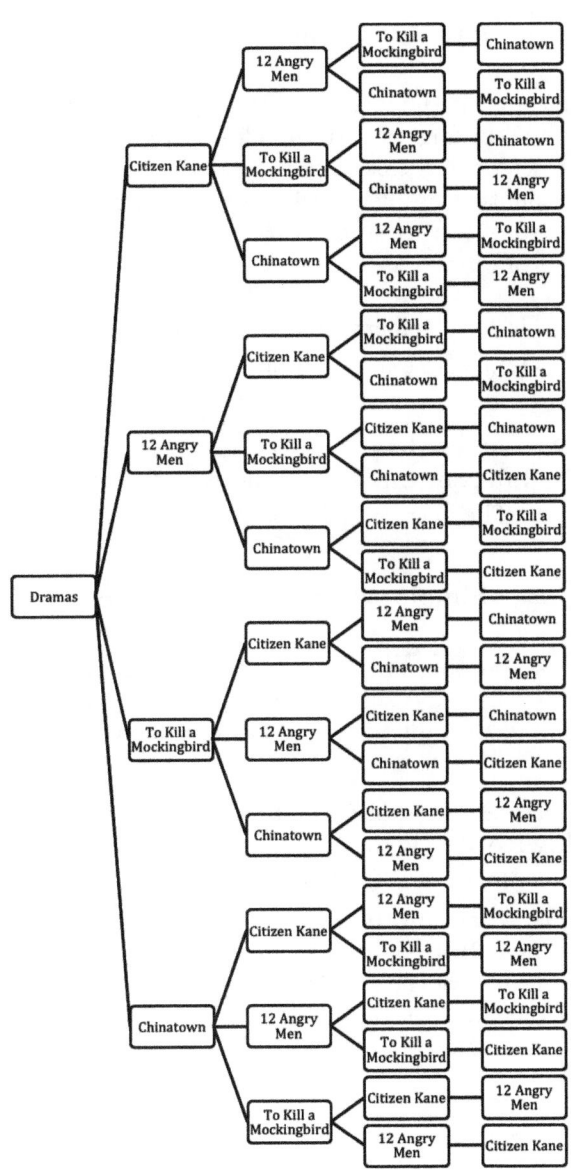

2.1 Basic Combinatorics Principles

There are two basic principles fundamental to counting.

FORMULA 2.1 **Multiplication Principle of Counting (MPC).**

When counting the total number of arrangements of certain objects according to certain rules, if there are n distinct ways of reducing this original counting problem to some other counting problem (called the "reduced problem") so that each of the reduced problems has the same rules and number of objects as any other reduced problem, then

number of arrangements of original problem
$= n \cdot$ (number of arrangements of any of the reduced problems)

EXAMPLE 2.2 **Sam has 7 shirts and 12 pants in his closet. How many different outfits can he wear?**
Solution: The answer would be $\underline{7} \cdot \underline{12} = 84$ by the MPC.[1]

FORMULA 2.2 **Simple Addition Principle (SAP).**

When a counting problem can be subdivided into two or more disjoint subcases, then the answer to the problem is the sum of the number of those two or more cases.

EXAMPLE 2.3 **How many triangles can be formed using the edges of Figure 1 below?**

Solution: There are 10 of the smaller size and 2 of the bigger size (formed from 4 of the smaller ones). So there are a total of $10 + 2 = 12$ triangles in this figure.

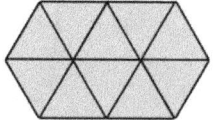

Figure 1

---

[1] We always assume there are no restrictions unless specifically listed. So any shirt can go with any pants, at least in theory. Even plaids with stripes.

2.2 Permutations

DEFINITION 2.1 **Permutation.**

A permutation or ranking of n objects is a listing of them in a certain order from first to last.

We must distinguish between permutations where repetition of objects is allowed (or, to put it another way, whether the objects are all distinct or not) and those where repetition is not allowed. A problem also comes up when there are duplicate objects being permuted.

FORMULA 2.3 **Permutations without Repetition.**

The number of permutations of length k from n distinct objects where repetition is not allowed can be expressed as follows:

$$_nP_k = n \cdot (n-1) \cdot (n-2) \cdots (n-k+1).[2]$$

EXAMPLE 2.4 **In the new Eastern European country of Mathuania, an official flag has yet to be created.** Mathuania's Federal Flag Commission has access to five different colors of fabric to create the country's new flag, and they want their new flag to be three different colored rectangles arranged horizontally. How many different flag choices does the FFC have?

Solution: The flag is really a permutation of the 5 colors. The "length" in this case is 3 because there are only three rectangles in the flag. There's no repetition because the FFC wants three different colors. Applying the formula of permutation we get $_5P_3 = 5 \cdot 4 \cdot 3 = 60$ different flag choices.

FORMULA 2.4 **Permutation with Repetition.**

The number of permutations of length k from n distinct objects where repetition is allowed can be expressed as follows:

$$n^k.$$

---

[2] Note: If all objects are permuted, the product continues until 1, and $_nP_n = n!$. Such was the case of the introductory example problem.

EXAMPLE 2.5 I have six letters to be delivered, and three boys (or girls) offer their services to deliver them. In how many ways have I the choice of sending the letters?[3]

Solution: One would naturally approach the problem as the boy delivering the letters; however, the problem is easier to solve from the perspective of the letters themselves. There are 3 choices for each letter as to which boy will deliver it. This is a case of permutation with repetition. There are 3 "distinct objects" (boys) each of 6 letters can chose from (the "length"). The solution, therefore, is $3^6 = 729$ ways.

FORMULA 2.5 **Permutation with Duplicate Objects.**

The number of permutations of a multiset of n objects made up of k distinct objects can be expressed as follows:

$$\binom{n}{n_1, n_2, \ldots n_k} = \frac{n!}{n_1! n_2! \cdots n_k!}$$

where n_i represents the multiplicity (frequency) of a distinct object i in the multiset.

A classic example is presented:

EXAMPLE 2.6 **How many ways can the letters of *Mississippi* be arranged?**

Solution: There are 11 letters but only 4 distinct letters. Specifically, there is 1 m, 4 i's, 4 s's, and 2 p's. These are the "multiplicities" of each of the distinct objects. Using the formula above, we get

$$\frac{11!}{1! \cdot 4! \cdot 4! \cdot 2!} = 34650$$

ways to arrange the letters in *Mississippi*.

2.3 Combinations

DEFINITION 2.2 **Combination.**

A **combination** of k objects taken from a collection of n objects is simply a selection of k of those distinct objects *without regard to order.*

---

[3]From *Choice and Chance*, W. A. Whitworth

2.3. COMBINATIONS

So "combination" in the mathematical sense does not mean exactly the same as a "locker combination" because in the mathematical sense the word combination does not care about the order of the elements chosen.

FORMULA 2.6 **Combinations without Repetition.**

The number of different combinations of k objects taken from a collection of n distinct objects (without repetition) is denoted by $\binom{n}{k}$ or $_nC_k$. A formula for calculating this number is:

$$\binom{n}{k} = {_nC_k} = \frac{n \cdot (n-1) \cdot (n-2) \cdots (n-k+1)}{k(k-1)(k-2) \cdots 1}$$

Before we say why this is true we give a practical example.

EXAMPLE 2.7 **How many different five man basketball teams can be chosen from a squad of eight players? (Here, a team is considered the same as long as it has the same players, regardless of who plays which position.)**

Solution: In this example, $n = 8$ and $k = 5$ such that the answer is $\binom{8}{5}$. So we compute it as follows:

$$\binom{8}{5} = \frac{8 \cdot 7 \cdot 6 \cdot 5 \cdot 4}{5 \cdot 4 \cdot 3 \cdot 2 \cdot 1} = 8 \cdot 7 = 56.$$

So there are 56 different teams which can be chosen.

Now the formula for $\binom{n}{k}$ may be proven as follows. Consider the problem of how many permutations of k objects there are where the k may be drawn without repetition from a collection of n objects. On the one hand by the same reasoning used above when discussing permutation, the number must be $n(n-1)(n-2)\cdots(n-k+1)$ since there are n choices for the first element, $n-1$ for the second, and so on until $n-k+1$ choices for the kth and final element of the permutation. On the other hand one could first choose k elements out of the n without regard to order, and *then* order them in $k!$ ways. By the MPC, this would give $\binom{n}{k} \cdot k!$ ways of permuting k objects drawn from n. These two ways of computing the number of such permutations must give the same answer so that

$$\binom{n}{k} \cdot k! = n(n-1)(n-2)\cdots(n-k+1)$$

Dividing both sides by $k!$ proves the result.

Calculations of combinations may sometimes be sped up by using alternate expressions for $\binom{n}{k}$.

FORMULA 2.7 **Alternate Expressions of the Basic Combination.**

$$\binom{n}{k} = \binom{n}{n-k}$$
$$\binom{n}{k} = \frac{n!}{k!(n-k)!}$$

The second of these is mainly useful if you have a calculator with a factorial button.

2.3.1 Combinations with Repetition: "Dashes and Dividers"

Now we turn to a most interesting situation: how to count the number of combinations of length n using objects of k different kinds (with an unlimited supply of each kind of object assumed). The solution is found by finding the number of permutations of n dashes and $k-1$ dividers. The number of dashes before the first divider corresponds to the number of objects of the first kind chosen; the number of dashes between the first and second divider corresponds to the number of objects of the second kind chosen, etc. This can be evaluated by selecting the positions of the dividers using the formula for combinations. There will be $n + k - 1$ positions total and $k - 1$ dividers. Using the formula we get the solution to be $\binom{n+k-1}{k-1} = \binom{n+k-1}{n}$. W.A. Whitworth, the father of modern combinatorics and author of *Choice and Chance*, denotes this as $_nR_k$.

FORMULA 2.8 **Combinations with Repetition (AKA "Dashes and Dividers" Formula).**

The number of Combinations of length n using k different kinds of objects is

$$_nR_k = \binom{n+k-1}{k-1} = \binom{n+k-1}{n} = \frac{(n+k-1)!}{n!(k-1)!}$$

2.3. COMBINATIONS

EXAMPLE 2.8 In how many ways can a purchaser select half a dozen handkerchiefs at a shop where seven sorts are kept?[4]

Solution: Use the "Dashes and Dividers" method to visualize this problem. There will be 6 handkerchiefs chosen, represented by 6 dashes.

— — — — — —

Then there will be $7 - 1 = 6$ dividers. Each divider also creates a space under it. An example being

In the above example, there is one of the first kind, zero of the second kind, one of the third kind, two of the fourth kind, zero of the fifth kind, two of the sixth kind, and zero of the seventh kind.

Thus, using the "Dashes and Dividers" formula there are $\binom{12}{6} = 924$ different selections to be made.

Here are some more examples of "Dashes and Dividers", also from *Choice and Chance* by W.A. Whitworth:

EXAMPLE 2.9 **A wood [forest] is full of primroses, violets, anemones, and bluebells. In how many ways can we compose a nosegay [bouquet] of 15 flowers?**

Solution: There are four different types of flowers to put in the bouquet (3 "dividers"). In the bouquet there are to be 15 flowers. The solution, using the formula, is $_{15}R_4 = \binom{18}{3} = 816$ different bouquets.

EXAMPLE 2.10 **At a post-office they keep ten sorts of postage-stamps. 1) In how many ways can a person buy twelve stamps? 2) In how many ways can he buy eight stamps? 3) In how many ways can he buy eight different stamps?**

Solutions:

1. The person is buying 12 stamps from 10 different types. There are 12 "dashes" and $10 - 1 = 9$ "dividers." Using the formula above there are $\binom{21}{9} = 293930$ ways. Choose wisely!

2. This person is buying 8 stamps from 10 different types. There are 8 "dashes" and again $10 - 1 = 9$ "dividers." Using the formula above there are $\binom{17}{9} = 24310$ ways.

---

[4]From *Choice and Chance*, W.A. Whitworth

3. We don't need "dashes and dividers" to answer this question. All you have to notice is that the question is *really* asking about a simple combination where the number of objects is 10 and the "length" is 8. Therefore the answer is $\binom{10}{8} = \binom{10}{2} = 45$ different selections.(Note the use of the simpler combination which is easier to calculate.)

The "Dashes and Dividers" formula can be used to prove the following:

FORMULA 2.9   **Number of Non-negative Integer Solutions.**

The number of solutions of the equation

$$x_1 + x_2 + \cdots + x_k = n$$

in non-negative integers is $\binom{n+k-1}{k-1}$. Note: x_1 and x_2, etc. are different entities, and order matters, such that, when using this formula, $7 + 2 = 9$ is different than $2 + 7 = 9$.

Whitworth sees this in a more practical light:

The number of ways in which n indifferent things can be distributed into k different parcels (empty parcels being admissible) is

$$\binom{n+k-1}{k-1}.$$

FORMULA 2.10   **Number of Positive Integer Solutions.**

The number of solutions of the equation

$$x_1 + x_2 + \cdots + x_k = n$$

in positive integers is $\binom{n-1}{k-1}$. Note: again x_1 and x_2, etc. are different entities, and order matters, such that, when using this formula, $7 + 2 = 9$ is different than $2 + 7 = 9$.

Again, This algebraic equation is related to a more practical example for Whitworth:

The number of ways in which n similar things can be distributed into k different parcels (empty parcels being inadmissible) is

$$\binom{n-1}{k-1}.$$

2.3. COMBINATIONS

We can consider why this is true by visualizing the selection of the separation of parcels with dashes and carrots:

The dashes above indicate the n similar things and the carrots will be where we separate the parcels. There are $n-1$ carrots, and to make k parcels, we only need to choose $k-1$ of them.

2.3.2 Other Interpretations of $\binom{n}{k}$

- Since order is irrelevant when specifying the members of a set it follows that $\binom{n}{k}$ is also equal to the total number of subsets of size k which exist in a set of size n.[5]

- The combination $\binom{n}{k}$ also gives the number of different "words" of length n which can be formed with a two-letter alphabet, there being k letters all the same, and $n-k$ other letters (of another symbol) all the same.[6]

EXAMPLE 2.11 **How many different seven letter "words" can be formed from the letters $\{u, u, u, u, u, v, v\}$? Examples of such "words" include uvuuuvu, vuuvuuu, and uvuvuuu.**

Solution: Note that once the positions of the two v's are chosen the word is determined. So this is really the problem of choosing two numbers (the positions where the v's will go) out of seven, *without regard to order*, since the same letter is going in each slot. Thus the answer is $\binom{7}{2} = 42/2 = 21$. There are 21 such "words".

Many counting problems seemingly unrelated to counting strange 'words' may in fact be essentially the same. Consider the following geometric problem:

EXAMPLE 2.12 **How many different paths are there from A to B in Figure 2 if one can only move up or to the right?**

Figure 2

---

[5] The total number of subsets, proper and improper, can be expressed as 2^n given that in a subset, a member can either be present or not. Note: one of these will be the empty set and another will be the set itself.

[6] With unlimited letters of each, the number of words would be equal to simply 2^n, or generalized to any alphabet of m letters, m^n.

Solution: To each path associate a "word" made up of 9 letters four of which must be u and five of which must be r. For example the path which goes four steps up and then five steps to the right is called 'uuuurrrrrr'. Since this sets up a one-to-one correspondence between the paths and such 'words' then there are $\binom{9}{5} = \binom{9}{4} = \frac{9 \cdot 8 \cdot 7 \cdot 6}{4 \cdot 3 \cdot 2 \cdot 1} = 9 \cdot 14 = 126$ such paths.

2.4 Recurrence Relations

DEFINITION 2.3   **Recurrence Relation.**

A recurrence relation is a formula or rule by which each term of a sequence (beyond a certain point) can be determined using one or more of the earlier terms.

Perhaps the most famous recurrence relation is the Fibonacci sequence (see Section 4.4).

In application to Combinatorics, it is best to give an example.

EXAMPLE 2.13   **Having only 1¢, 2¢, and 5¢ stamps, how many ways can any number of these be arranged in a row so that they add up to n¢ (call it a_n) where $n \geq 6$?**[7]

Solution: By listing the first few n's we will hopefully see a pattern.

n¢	$1 + \cdots$	$2 + \cdots$	$5 + \cdots$	Total a_n
1¢	1			1
2¢	$1+1$	2		2
3¢	$1+1+1$ $1+2$	$2+1$		3
4¢	$1+1+1+1$ $1+1+2$ $1+2+1$	$2+1+1$ $2+2$		5
5¢	$1+1+1+1+1$ $1+1+1+2$ $1+1+2+1$ $1+2+1+1$ $1+2+2$	$2+1+1+1$ $2+1+2$ $2+2+1$	5	9

---

[7]From *Combinatorics: A Problem Oriented Approach* by Daniel A. Marcus

2.5. PIGEONHOLE PRINCIPLES

We can definitely see a recurrence pattern forming. Take for example 4¢. The number of $1 + \cdots$ is the same as the total number of 3¢. And the number of $2 + \cdots$ is the same as the total number of 2¢. We see that for 5¢, the number of $1 + \cdots$ is the same as the total number of 4¢, and the number of $2 + \cdots$ is the same as the total number of 3¢. Why? Because for n¢ we can start with $1 + \cdots$ and to add on the other $(n-1)$¢ there are the same number of ways $(n-1)$¢ can be made. Similarly we can start with $2 + \cdots$ and to add the other $(n-2)$¢ there are the same number of ways $(n-2)$¢ can be made. We can extrapolate this, i.e. if we start with $5 + \cdots$, to add the other $(n-5)$¢ there are the same number of ways $(n-5)$¢ can be made.

Therefore, we have discovered the recurrence relation:

$$a_n = a_{n-1} + a_{n-2} + a_{n-5}.$$

2.5 Pigeonhole Principles

Now consider a famous principle first clearly stated by the German mathematician Dirichlet:

FORMULA 2.11   **The Pigeonhole Principle**

> If m pigeons must be put in n pigeonholes, and $m > n$, then there must be at least one pigeonhole holding more than one pigeon.

FORMULA 2.12   **The Pigeonhole Principle—Refined Version**

> If at least $nk + 1$ objects must be put in n boxes then there must be at least one box holding at least $k + 1$ objects.

Though the Pigeonhole Principle is fundamental, contest examples are rare. Here are a few examples that demonstrate Dirichlet's principle:

EXAMPLE 2.14   **Dave is a messy college student, and today he's in a rush. In his unorganized dresser drawer are four individual black socks, six individual blue socks, and two individual white socks. 1) What's the fewest number of socks Dave should pull out to guarantee he can make a pair of any color? 2) What's the fewest number of socks Dave should** pull out to guarantee he gets a pair of black socks for his job interview?

Solutions:

1. To answer these pigeonhole questions, always imagine the worst possible situation. Here it is: Dave pulls out a black sock, then a blue sock, then a white sock. He can't make a pair! But the next sock that he pulls out must be either black, blue, or white. Therefore, after his fourth pull, Dave has a pair! The answer is 4.

2. Again, think of the worst possible situation: Dave pulls out both white socks and then all six blue socks (8 pulls) before pulling out his first black sock. All that's left in the drawer then is three black socks. On his next pull, he'll get a black sock to make a pair. All together that's nine pulls unsuccessful and on the tenth pull he'll get a pair. The answer is 10.

2.6 How to Avoid Over Counting

Let $|A|$ equal the number of elements in the finite set A. Suppose you know $|A|$ and $|B|$. Does that mean you know $|A \cup B|$? Not necessarily, because A and B may or may not have elements in common. To avoid counting such common elements twice use P. I. E.

FORMULA 2.13   **Principle of Inclusion-Exclusion (P. I. E.) for Two Sets.**

> The size of the union of two finite sets can be expressed as the sum of the sizes of each set less the size of their intersection. So that:
>
> $$|A \cup B| = |A| + |B| - |A \cap B|$$

This is true because in $A \cup B$ there is overlap (i.e. $A \cap B$) such that when you count the members in A and those in B, you *double count* members that are in both sets. That's why you subtract out $|A \cap B|$.

Note: If A and B are disjoint then the last term of this formula vanishes and the number of elements of the union does, in this case only, reduce to the sum of the sizes of A and B.

In the same principle for three sets, the number of terms is increased greatly.

2.6. HOW TO AVOID OVER COUNTING

FORMULA 2.14 **P. I. E. for Three Sets.**

The size of the union of three finite sets can be expressed as the sum of the sizes of each set, less the size of the intersections of any two sets, plus the size of the intersection of all three sets. (Fig. 3) So that:

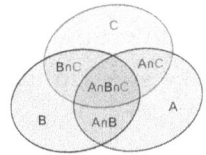

Figure 3

$$|A \cup B \cup C| = |A| + |B| + |C|$$
$$-|A \cap B| - |B \cap C| - |A \cap C|$$
$$+|A \cap B \cap C|$$

This is true because in $A \cup B \cup C$ there is overlap (i.e. $A \cap B$, $B \cap C$, $A \cap C$, and $A \cap B \cap C$) such that when you count the members in A, those in B, and those in C, you *double count* members that are in two of the sets. That's why you subtract out $|A \cap B|, |A \cap C|$, and $|B \cap C|$. When you do that, you end up subtracting out members of $A \cap B \cap C$ as well; therefore you have to add those back!

This formula can be generalized for any number of sets. Remember to always add the size of intersections of odd numbers of sets and to subtract the size of intersections of even numbers of sets.

We give some examples to help explain.

EXAMPLE 2.15 **Given as set $\mathcal{S}$ containing the first one hundred natural numbers, how many of those numbers are multiples 6 or 7?**

Solution: Note $\mathcal{S} = \{1, 2, \ldots 100\}$. Of those 100 integers, one out of every six is a multiple of 6 and one out of every seven is a multiple of 7. That means that there are $\lfloor 100/6 \rfloor = \lfloor 16\frac{2}{3} \rfloor = 16$ multiples of 6 in $\mathcal{S}$. (Note $\lfloor n \rfloor$ is the floor function, which rounds n down to the nearest whole integer.) Similarly there are $\lfloor 100/7 \rfloor = \lfloor 14\frac{2}{7} \rfloor = 14$ multiples of 7 in $\mathcal{S}$. But the answer isn't $16 + 14$. We have not accounted for the fact that we've double counted the multiples of both 6 and 7, i.e. the multiples of 42. There are $\lfloor 100/42 \rfloor = \lfloor 2\frac{8}{21} \rfloor = 2$ multiples of 42. Using P. I. E., there are $16 + 14 - 2 = 28$ multiples of 6 or 7 in $\mathcal{S}$.

EXAMPLE 2.16 **In a survey on the chewing gum preferences of baseball players, it was found that**

- 22 like fruit.
- 25 like spearmint.
- 39 like grape.
- 9 like spearmint and fruit.
- 17 like fruit and grape.
- 20 like spearmint and grape.
- 6 like all flavors.
- 4 like none.

How many players were surveyed?

Solution: This is a simple example of P. I. E. using three sets: fruit gum chewers (F), spearmint gum chewers (S), and grape gum chewers (G). Therefore, using P. I. E., substitute for the values given above such that

$$\begin{aligned} |F \cup S \cup G| &= |F| + |S| + |G| \\ &\quad - |F \cap S| - |F \cap G| - |S \cap G| \\ &\quad + |F \cap S \cap G| \\ |F \cup S \cup G| &= 22 + 25 + 39 \\ &\quad -9 - 17 - 20 \\ &\quad +6 \\ &= 46 \end{aligned}$$

46 players who like those flavors. But don't forget about the 4 who didn't like any of them! Add those to 46 to get $4 + 46 = 50$ total players surveyed.

2.7 Multifaceted Examples in Combinatorics

Here we present examples in combinatorics that require many of the principles listed above. We will try to identify each principle as we use it.

EXAMPLE 2.17 A spider has one sock and one shoe for each of its eight legs. In how many different orders can the spider put on its socks and shoes, assuming that, on each leg, the sock must be put on before the shoe?[8]

---
[8]From AMC12 2001

2.7. MULTIFACETED EXAMPLES IN COMBINATORICS

Solution: Number the spiders' legs 1 through 8, and let a_k denote the sock that will go on leg k and b_k denote the shoe that will go on leg k. Without the restriction that the sock has to go on before the shoe, this problem would be a simple permutation where each of 16 things is permuted, the result being 16!. But the restriction says that the sock has to go on before the shoe. What does this mean for our permutation? It means that b_1 cannot come before a_1 and that b_2 cannot come before a_2, etc. In the 16! permutations, b_1 comes before a_1 in exactly half of them, so we can divide 16! by 2, leaving us with $16!/2$ permutations. In these remaining permutations, b_2 comes before a_2 again in exactly half of them, so we can divide $16!/2$ by 2, leaving us with $16!/2^2$. Again, this is true for every k from 1 to 8. So, following the pattern, by the time we get to b_8 coming before a_8, we will divide the number of remaining permutations by 2 to get $16!/2^8$. Therefore, there are $16!/2^8 = 81\ 729\ 648\ 000$ different orders the spider can put his 8 socks and 8 shoes on his feet.[9]

The following examples feature casework:

EXAMPLE 2.18 **In how many ways can the letters in *quisquis*[10] be arranged without letting two letters which are alike come together.**[11]

Solution: Like most problems with case work, start by finding the unrestricted possibilities, i.e., the number of ways to arrange *quisquis*. This is done using the permutation with duplicates formula. There are 8 letters, 2 of each, so the total unrestricted arrangements is $\binom{8}{2,2,2,2} = \frac{8!}{2!^4} = 2520$ arrangements. Unfortunately, this includes arrangements like *qquissiu*, where some letters come together. To counteract this, we need to dive into casework to find the number of arrangements in violation of the restriction.

Let's do some prep before we dive into case work. First we need to notice that what's true for one letter, is true for any letter. That is, for example, when we talk about only the q's being together, the number we find must be multiplied for each letter which could be together. Let's call this the "frequency" of the case, so that in the end, we multiply the frequency by the size of the case to find the total number of violations in that case. Second, let's set up the notation $\boxed{qq}$ to mean the two q's are "glued together," that is, they are one entity. Third, don't forget to account for double counting. You'll see how this comes up later.

---

[9] Solution from *102 Combinatorical Problems* by Titu Andrescu and Zuming Feng
[10] For those curious, *quisquis* is Latin, meaning whoever or whatever.
[11] From *Choice and Chance*, W. A. Whitworth, Exercise #97.

The first case is that where all the letter pairs are "glued together", that is we have a word like $\boxed{qq}\boxed{uu}\boxed{ii}\boxed{ss}$. Each box is its own entity, therefore, the number of arrangements is the result a simple permutation: $4! = 24$ arrangements. The frequency of this case is only 1, because there's only one way to choose all 4 pairs to be glued together.

The second case is that where 3 of the letter pairs are glued together, that is we have a word like $\boxed{qq}\boxed{uu}s\boxed{ii}s$. In this case, the s's are not glued together, therefore, there are 5 distinct entities. Using the permutation with duplicates formula, then there are $\binom{5}{1,1,1,2} = \frac{5!}{2} = 60$ arrangements like this. However, in some of these cases, the s's are together! We've double counted! Counteract the double counting by subtracting from 60 the 24 arrangements where the pairs are all glued together. That's 36 arrangements where only 3 of the letter pairs are glued together. The frequency of this case is 4 because there are 4 letter pairs and we're choosing 3 to be glued together, and $\binom{4}{3} = 4$.

The third case is that where 2 of the letter pairs are glued together, that is we have a word like $\boxed{qq}\boxed{uu}sisi$. In this case, the s's and the i's are not glued together, therefore, there are 6 distinct entities and $\binom{6}{1,1,2,2} = \frac{6!}{2^2} = 180$ arrangements. However, in some of these cases, the s's are together, in others the i's are together, and in still others, both the s's and the i's are together! We've double counted! Counteract the double counting by subtracting from 180 the 24 arrangements where the pairs are all glued together, and twice the 36 arrangements, accounting for both cases where we double counted the arrangements where only the s's are glued together and where only the i's are glued together. That's $180 - 24 - 2(36) = 84$ arrangements where only 2 of the letter pairs are glued together. The frequency of this case is 6 because there are 4 letter pairs and we're choosing 2 to be glued together, and $\binom{4}{2} = 6$.

The last case is that where 1 of the letter pairs are glued together, that is we have a word like $\boxed{qq}iususi$. Following the previous examples there are $\binom{7}{1,2,2,2} = \frac{7!}{2^3} = 630$ arrangements. But like in the previous cases, we've double counted! Counteract the double counting by subtracting from 630 (1) the 24 arrangements where the pairs are all glued together, (2) three times the 36 arrangements, accounting for the three cases where we double counted the arrangements where one of the other letter pairs is glued together, and (3) three times ($\binom{3}{2}$) the 84 arrangements, accounting for the three cases where we double counted the arrangements where two of the other letter pairs is glued together. That's $630 - 24 - 3(36) - 3(84) = 246$ arrangements where only 1 of the letter pairs are glued together. The frequency of this case is 4 because there are 4 letter pairs and we're choosing 1 to be glued together, and $\binom{4}{1} = 4$.

To find the final answer, we take the original 2520 arrangements and subtract out the violations. For each case, don't forget to multiply by the

2.7. MULTIFACETED EXAMPLES IN COMBINATORICS

frequency!

$$2520 - 24 - 4(36) - 6(84) - 4(246)$$
$$= 864 \text{ arrangements where no two letters come together}$$

EXAMPLE 2.19   In how many ways can five black balls, five red balls, and five white balls be distributed into three different bags [Bag_1, Bag_2, Bag_3], five into each?[12]

Solution: This problem is especially difficult because this question does not fit into any one of the methods used earlier. There are definitely deep "dashes and dividers" roots in this question but unlike those examples, there is a limited number of each color. We will have to find the number of ways in the unrestricted case, then subtract out those ways that violate the restrictions. This "whittling away" to match the restrictions is a fundamental technique in counting.

Start by noticing that if we choose which balls to put in the first two bags, we will predetermine which balls will go in the last bag. Therefore, we shall concentrate our efforts in counting the ways to put 5 balls into each of the first two bags.

The first bag has a "dashes and dividers" selection. It has 5 "dashes" and $3 - 1 = 2$ "dividers." Using the formula, we have

$$Bag_1 = \binom{7}{2} = 21 \text{ different selections.}$$

Similarly the second bag has the same number of selections, i.e. $Bag_2 = 21$. Using the MPC, therefore, we can say that there are $21 \cdot 21 = 441$ unrestricted selections.

The issue to address is that some of these selections violate the restriction on the number of balls of each color. For instance, there exists a selection where Bag_1 has 4 black balls and 1 white ball, and Bag_2 has 3 black balls, 1 white ball, and 1 red ball. That's seven black balls, when there are only five! We must counteract the over counting by subtracting out the bad cases. Before doing this, we should do a little prep work. First, note that what's true for one color ball, is true for each other color; that is, we can just subtract out three times what we find to be in violation for one color. Second, in case work, be careful to only change one variable overall. To be safe, don't try to fit everything under one umbrella mathematical expression. On the other hand, be efficient and see how one case can be expressed succinctly. In this kind of case work, you play the part of "Math Inspector." You're summing up the number of selections "in violation" of the parameters of the question and then subtracting the sum from the number of unrestricted cases.

---

[12]From *Choice and Chance*, W. A. Whitworth, Exercise #199.

So let's just work with the case where the black colored balls are in violation. Furthermore, let's only vary the number of black balls Bag_1 has to create each case. Start the case work with Bag_1 having the maximum of 5 black balls, drawn as below. There's only 1 Bag_1 where this is the case, because the remaining one "divider" only has one place to choose. Now we need to find the corresponding Bag_2's where too many black balls will be chosen. In fact *any number* of black balls will cause an excess. So, as long as we ensure one black ball is in Bag_2, there will be an excess. The two dividers in Bag_2 can go anywhere. There are 6 remaining places to put the two dividers, making $\binom{6}{2} = 15$ ways for Bag_2 to be in violation given Bag_1 has 5 black balls. We visualize this case below:

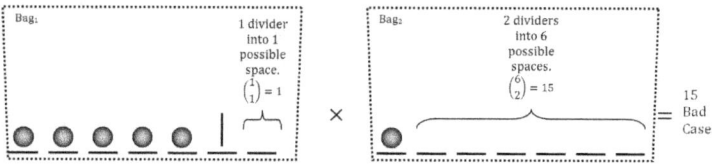

In the next case we look at what happens when Bag_1 has 4 black balls. In this case the other "divider" has now a choice of two places: two Bag_1's with 4 black balls. For this to be an invalid selection, Bag_2 has to have at least 2 black balls. That means the two dividers can be placed in any of the last five spaces, i.e. there are $\binom{5}{2} = 10$ ways for the selection to be in violation given Bag_1 has 4 black balls. Using the MPC, the total number of ways for this kind of selection to be in violation is the product of the ways Bag_1 can have 4 black balls and for Bag_2 to have 2 or more black balls. Therefore, the total number of ways is $2 \cdot 10 = 20$ ways in the case where Bag_1 has 4 black balls. We visualize again below.

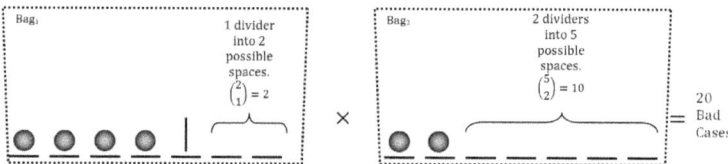

We can see a pattern forming: in each case we have a combination with length 1 multiplied by a combination with length 2. And as the number of objects of the first combination goes up by one, the number of objects in the second combination goes down by one. We can verify that in the last case where Bag_1 has only 1 black ball, there will be $\binom{1}{1}\binom{2}{2} = 5$ ways to have an excess of black balls. We can extrapolate the pattern, and add all the cases together using the SAP, to find the total number of selections with an excess

2.7. MULTIFACETED EXAMPLES IN COMBINATORICS

of black colored balls to be

$$\binom{1}{1}\binom{6}{2} + \binom{2}{1}\binom{5}{2} + \binom{3}{1}\binom{4}{2} + \binom{4}{1}\binom{3}{2} + \binom{5}{1}\binom{2}{2}$$
$$= 1 \cdot 15 + 2 \cdot 10 + 3 \cdot 6 + 4 \cdot 3 + 5 \cdot 1 = 70 \text{ ways}$$

Furthermore, there are three colors balls; therefore, the total number of violations is $3 \cdot 70 = 210$ violations. It is safe to say this because the way we set up the casework was such that only one color was in violation at a time. In other words, because there is no way for two colors to be in violation simultaneously, we don't need to worry about inclusion/exclusion.

To answer the question posed, we take the original 441 unrestricted selections and subtract away the violations. Therefore, there are $441 - 210 = 231$ ways five black balls, five red balls, and five white balls can be distributed into three different bags, five into each bag.

Chapter 3

Probability and Statistics

3.1 The Basics of Probability

DEFINITION 3.1   Event A and Complementary Event A'.

An event A is a set of outcomes to which a probability is assigned. The complementary event A' is the set of all other outcomes.

NOTATION 3.1   The Probability of an Event Occurring.

The notation $p(A)$ means the probability of the event A occurring. Similarly, $p(A')$ means the probability of the event A not occurring.

AXIOM 3.1   The Unity of Probability.

Given a probability $p(A)$, it is always true that

$$0 \leq p(A) \leq 1,$$

and it is always true that

$$p(A) + p(A') = 1.$$

The second part is rather intuitive. It is reasonable to be sure that something will either happen or it won't.

Axiom 3.2 The Ways an Event May Occur.

If an event A may occur in many different disjoint ways, i.e. $A_1, A_2, A_3, \ldots, A_k$, then it is always true that,

$$p(A) = p(A_1) + p(A_2) + p(A_3) + \cdots + p(A_k)$$

Definition 3.2 Odds.

If it is said that the odds of event A are r to q in favor of an event, then
$$p(A) = \frac{r}{r+q}$$
and
$$p(A') = \frac{q}{r+q}.$$

Similarly, if it is said that the odds of event A are r to q against an event, then

$$p(A) = \frac{q}{r+q}$$
and
$$p(A') = \frac{r}{r+q}.$$

Definition 3.3 Probability.

If there be a number of events, r, of which all are likely, and if any one of a number of events, q, will produce a certain result, A, that will otherwise not happen, the probability that A will occur can be expressed as

$$p(A) = \frac{q}{r}.$$

This definition is at the foundation of probability. It seems very intuitive, but it is nonetheless crucial to understanding probability. Often these q's and r's will be the result of a combinatorics problem. For Example:

3.1. THE BASICS OF PROBABILITY

EXAMPLE 3.1 Three balls are to be drawn [without replacement] from an urn which contains five black, three red, and two white balls. What is the chance of drawing two black balls and one red?[1]

Solution: The total number of ways to draw can be expressed as a combination: $\binom{10}{3} = 120$ ways, all equally likely. Now, two black balls can be selected in $\binom{5}{2} = 10$ ways, and the red ball can be chosen in three ways. By the MPC, the ways to choose 2 black balls and 1 red ball is $10 \cdot 3 = 30$. Using the definition of probability, the probability of choosing 2 black balls and 1 red ball is $\frac{30}{120} = \frac{1}{4}$.

FORMULA 3.1 **Expectation Value.**

Given a finite number of events, $(A_1, A_2, A_3, \ldots, A_n)$, the occurrence of each having a quantitative outcome, $(a_1, a_2, a_3, \ldots, a_n)$, the Expectation Value, $E(A)$, equals the sum of the products of the probability of each event with its quantitative outcome. In other words:

$$E(A) = \sum_{k=1}^{n} a_k p(A_k)$$
$$= a_1 p(A_1) + a_2 p(A_2) + a_3 p(A_3) + \cdots + a_n p(A_n)$$

EXAMPLE 3.2 In the Sohcahtoa Casino, found in Longleg, OK, is a standard roulette wheel with a 0, a 00, and the numbers 1 through 36. (1) Chinchy Chad places a $1 bet on the "six line" of the numbers 25 through 30, which pays 5-to-1 (that is, if the roulette ball falls into the pocket of one of these numbers, the payout will be 5 times his bet). What is Chad's expectation value? (2) Risky Rachel places a $20 bet straight-up on the number 15, which pays 35-to-1, and $50 on the odds, which pays 1-to-1. What is Rachel's expectation value?

Solutions:

1. In Chad's bet, there is a $\frac{6}{38}$ chance that he wins $5 and a $\frac{32}{38}$ chance that he loses $1. Therefore, the expectation value is

$$(\$5)\frac{6}{38} + (-\$1)\frac{32}{38} = \$\frac{30-32}{38} = -\$\frac{2}{38} \approx -\$.053$$

---
[1] From *Choice and Chance*, W. A. Whitworth

2. In Rachel's bet, there is a $\frac{1}{38}$ chance that the ball falls into pocket 15 and she wins the ridiculous sum of $750. There is a $\frac{17}{38}$ chance that the ball falls into one of the other 17 odd pockets in which case she wins $50, but loses $20, making her net winnings $30. Then, of course there is a $1 - \frac{17}{38} - \frac{1}{38} = \frac{20}{38}$ chance that she loses all $70. Her expectation value is therefore

$$(\$750)\frac{1}{38} + (\$30)\frac{17}{38} + (-\$70)\frac{20}{38} = -\$\frac{140}{38} \approx -\$3.68$$

The house always wins.

3.2 Probability of Multiple Events

NOTATION 3.2 $p(A \cap B)$ and $p(A \cup B)$.

The notation $p(A \cap B)$ means the probability of the events A and B both occurring.

The notation $p(A \cup B)$ means the probability of at least one of the events A or B occurring.

DEFINITION 3.4   **Independent Events.**

Two events A and B are independent if the occurrence or non-occurrence of each does not affect the probability of the other occurring.

FORMULA 3.2   **Probability of Independent Events.**

Given independent events A and B, the probability that both A and B occur is

$$p(A \cap B) = p(A)p(B).$$

NOTATION 3.3   **Conditional Probability.**

The conditional probability of event B occurring given that event A has already occurred is denoted $p(B|A)$.

3.3. BASIC AND COMPLEX EXAMPLES IN PROBABILITY

FORMULA 3.3   **Probability of Dependent Events.**

Given two events A and B which are not independent, i.e. they are dependent, the probability that both A and B will occur is equal to the product of the chance that A should happen and the chance that when A happens, B should also happen.

In other words:
$$p(A \cap B) = p(A)p(B|A)$$

This formula is continuable such that given any number of events, A, B, C,..., Z the probability that all events should happen is

$$p(A \cap B \cap C \cap \cdots \cap Z) = p(A)p(B|A)p(C|B,A)\cdots p(Z|Y,X,\ldots,B,A).$$

3.3 Basic and Complex Examples in Probability

Here we give a wide variety of examples dealing with probability, most of which come from W. A. Whitworth's *Choice and Chance*.

EXAMPLE 3.3   **If a letter be taken at random from the word *organize*, what is the chance it is a vowel?**

Solution: This question needs a simple application of the definition of probability to be solved. There are 8 letters total, 4 of which are vowels, therefore, the probability is $\frac{4}{8} = \frac{1}{2}$.

EXAMPLE 3.4   **Three letters are taken at random from *association*, (1) what is the chance that one of them is a *c*? (2) What is the chance that 2 of them are alike?**

Solutions:

1. This question is not as simple. To do the combinatorics, pick the c first. Then, there are $\binom{10}{2} = 45$ choices for the other 2 letters. There are $\binom{11}{3} = 165$ ways to pick 3 letters overall. Therefore the chance that we pick a c is $\frac{45}{165} = \frac{3}{11}$.

2. This question is more complicated still. There are 165 ways to pick the 3 letters overall as we solved in the previous part of the solution. To pick a double letter, we either have to pick an *ss*, an *ii*, an *oo*, or an *aa*. For each of those 4 selections there are 9 other

choices for the remaining letter. Therefore, the total selections where a double letter occurs is $4 \cdot 9 = 36$ ways, and the probability is $\frac{36}{165} = \frac{12}{55}$.

EXAMPLE 3.5 A letter is taken at random out of each of the words *choice* and *chance*: what is the chance that they should be the same letter?

Solution: There are 6 choices for the selection out of *choice* and 6 for *chance* making 36 total selections. The possibilities of same letters are *cc*, *hh*, and *ee*. The duplicate *cc* happens $\frac{4}{36}$, because there are two ways to pick a *c* from *choice* and two ways to pick a *c* from *chance*, and $2 \cdot 2 = 4$. As for the other letters, *hh* happens $\frac{1}{36}$, and *ee* happens $\frac{1}{36}$, summing to $\frac{6}{36} = \frac{1}{6}$.

EXAMPLE 3.6 [Aaron and Moses] stand in a line with 10 other persons. If the arrangement is made at random, what is the chance that there are exactly 3 persons between [Aaron and Moses]?

Solution: Think of this in combinatorics terms. There are 12! arrangements for the line. Aaron and Moses need to be 3 places from each other, (i.e. $\underline{M}\ \underline{\ }\ \underline{\ }\ \underline{\ }\ \underline{A}\ \underline{\ }\ \underline{\ }\ \underline{\ }\ \underline{\ }\ \underline{\ }\ \underline{\ }\ \underline{\ }$) which happens 16 times the number of combinations the other people can stand, $16 \cdot 10!$ arrangements. (It happens 16 times because the block of 5 made with Aaron, Moses, and the 3 people in between can be placed in 8 different places in the line, and it can be inverted such that either Moses or Aaron is near the front of the line. $8 \cdot 2 = 16$.) The probability, therefore is $\frac{16 \cdot 10!}{12!} = \frac{16}{12 \cdot 11} = \frac{4}{33}$.

EXAMPLE 3.7 Two [positive integers] are chosen at random. Find the chance that their sum is even.

Solution: The only way to get an even sum is to sum two even positive integers or to sum two odd positive integers. Each integer has a $\frac{1}{2}$ chance of being even and a $\frac{1}{2}$ chance of being odd. Therefore, there is a $\frac{1}{2} \cdot \frac{1}{2} = \frac{1}{4}$ chance that both integers are even, and a $\frac{1}{2} \cdot \frac{1}{2} = \frac{1}{4}$ that both integers are odd. Therefore, the chance that two integers sum is even is $\frac{1}{4} + \frac{1}{4} = \frac{1}{2}$.

EXAMPLE 3.8 Out of a set of dominoes, numbered from double one to double six, one is drawn at random. At the same time, a pair of common dice are thrown. (1) What is the chance that the numbers turned up on the dice will be the

3.4. STATISTICS

same as those on the domino? (2) What is the chance that they will have one number at least in common?
Solutions:

1. The number of dominos is a "Dashes and Dividers" problem with 2 "dashes" (for the 2 numbers that will be displayed on the domino) and $6 - 1 = 5$ "dividers" (for the 6 different numbers that can be chosen). Therefore the number of dominos is $\binom{7}{2} = 21$ dominos. And there will only be 1 roll like this, so the probability is $\frac{1}{21}$.

2. This is a example of conditional probability. In the case where the dice throw results in a doublet (the chance of that being $\frac{1}{6}$) there are 6 dominos which will have the number of the doublet (the chance of picking one of those being $\frac{6}{21}$). In the other cases (the chance of those being $\frac{5}{6}$ by the Unity of Probability), there are 10 dominos with one number being the same and 1 with both being the same, a total of 11 dominos with at least 1 number in common (the chance of picking one of those being $\frac{11}{21}$). Therefore the total probability is

$$\frac{1}{6} \cdot \frac{6}{21} + \frac{5}{6} \cdot \frac{11}{21} = \frac{61}{126}.$$

EXAMPLE 3.9 **An archer hits his target on average 3 times out of 4. Find the chance that in the next four trials, he will hit it three times exactly.**
Solution: There are 4 ways in which ✓ ✓ ✓ ○ can occur. And the chance of each is $\frac{3}{4} \cdot \frac{3}{4} \cdot \frac{3}{4} \cdot \frac{1}{4}$. Therefore the total chance is 4 times that or $\frac{27}{64}$.

3.4 Statistics

3.4.1 Data Distribution

DEFINITION 3.5 **Range.**

> The range is represented by length of the smallest interval which contains all the data. It is obtained by subtracting the smallest value from the largest value.

DEFINITION 3.6 **Median.**

The median of a data set is the numeric value separating the higher half of a data sample from the lower half.

FORMULA 3.4 **Obtaining the Median.**

To obtain the median of a data set, arrange the data in an increasing list.

- If there are an odd number $(2n-1)$ of data points, the median is the middle (n^{th}) data point.
- If there are an even number $(2n)$ of data points, the median is equal to the arithmetic mean of the middle two $(n^{\text{th}}$ and $(n+1)^{\text{th}})$ data points.

DEFINITION 3.7 **Mode.**

The mode(s) of the data set reflect(s) the data point(s) which occur(s) most often.

EXAMPLE 3.10 **What is (are) the mode(s) of the data set $\{1, 2, 6, 6, 31, 72, 72, 73, 101\}$?**
Solution: There are two values which occur most often, 6 and 72. They are the two modes of this data set.

DEFINITION 3.8 **Mean.**

In mathematics a mean is a value intermediate between a multiplicity of values contained in a data set.

There are three kinds of means:

3.4.2 Three Kinds of Means

FORMULA 3.5 **Arithmetic Mean.**

The Arithmetic Mean is the simplest. The mean of the set $\{a_1, a_2, \ldots, a_n\}$ is the sum of the values, divided by the quantity of values summed.

$$\frac{a_1 + a_2 + \cdots + a_n}{n}$$

$= $ arithmetic mean of the data points a_1 through a_n

The arithmetic mean is the mean used most often in competition problems.

3.4. STATISTICS

FORMULA 3.6   **Geometric Mean.**

The Geometric Mean is calculated by taking the product of the n positive values of the set $\{a_1, a_2, \ldots, a_n\}$, taken to the nth root.

$\sqrt[n]{a_1 \cdot a_2 \cdots a_n}$
 $=$ geometric average of the data points a_1 through a_n

FORMULA 3.7   **Harmonic Mean.**

The Harmonic Mean is calculated by multiplying the number of non-zero data points of the set $\{a_1, a_2, \ldots, a_n\}$ by the reciprocal of the sum of the reciprocals of the data points.

$\dfrac{n}{\frac{1}{a_1} + \frac{1}{a_2} + \cdots + \frac{1}{a_n}}$
 $=$ harmonic average of the data points a_1 through a_n

Use the harmonic mean in these situations

- **Physics:** If a vehicle (person, train, car, bicycle, etc.) travels at x mph for a distance and then y mph for the same distance, the average speed is calculated by the harmonic mean. This can be extended to any number of speeds at the same distances. Average Speed $= \dfrac{2}{\frac{1}{x}+\frac{1}{y}}$ mph or $= \dfrac{n}{\frac{1}{x_1}+\frac{1}{x_2}+\cdots+\frac{1}{x_n}}$ mph.

- **Algebraic Work:** If it takes one person (machine, widget, etc.) x hours to do a job, and it takes another person y hours to do the same job, it takes both persons half the harmonic mean of the time it takes each person to do the job. If both work together, the job gets done in $\dfrac{1}{\frac{1}{x}+\frac{1}{y}}$ hours, half the harmonic mean of x and y. If many work together, the job gets done in $\dfrac{1}{\frac{1}{x_1}+\frac{1}{x_2}+\cdots+\frac{1}{x_n}}$ hours.

- **Geometry:** In the classic case of crossing wires, such as in the figure on the following page, where two right triangles with the same base and different heights (h_1 and h_2) are superimposed, the height of the point of intersection of the hypotenuses from the base is equal to $\dfrac{1}{\frac{1}{h_1}+\frac{1}{h_2}}$, which is half the harmonic mean of h_1 and h_2.

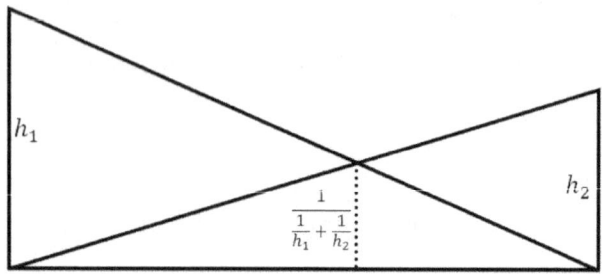

FORMULA 3.8 **The Geometric-Arithmetic-Harmonic Inequality.**

For any set of data $\{a_1, a_2, a_3, \ldots, a_n\}$,

$$\frac{n}{\frac{1}{a_1} + \frac{1}{a_2} + \cdots + \frac{1}{a_n}} \leq \sqrt[n]{a_1 \cdot a_2 \cdots a_n} \leq \frac{a_1 + a_2 + \cdots + a_n}{n}.$$

3.4.3 Examples in Statistics

EXAMPLE 3.11 Consider the weather data for 10 days shown in the table below. John identified the mode of the daily high temperature and the mode of the daily low temperature, in degrees Fahrenheit, based on this data. What is the mean of those two numbers?[2]

	M	Tu	W	Th	F	Sa	Su	M	Tu	W
$High$	72	74	70	69	72	74	78	79	72	70
Low	61	58	60	58	63	58	63	58	60	59

Solution: Remember, the mode is the number that most frequently occurs in a set of data. So the mode for the high is 72 because it occurs 3 times in that set of data. 74 and 70 both occur twice but 72 occurs 3 times so it is clearly the mode. The mode for the low is without doubt 58 because it occurs 4 times in that set of data.

The question asks about the mean of those two numbers. In this case, mean has to be interpreted in context as algebraic mean, that is, the average of the two modes. To find the average we simply add and divide by 2.

$$\frac{72 + 58}{2} = \frac{130}{2} = 65$$

The answer, therefore, is 65°F.

---

[2] From MATHCOUNTS 2014 School Competition, Sprint Round

3.4. STATISTICS

EXAMPLE 3.12   Nick is a runner, and his goal is to complete four laps around a circuit at an average speed of 10 mph. If he completes the first three laps at a constant speed of only 9 mph, what speed does he need to maintain in miles per hour on the fourth lap to achieve his goal?[3]

Solution: What we're trying to find is the missing piece to make the current average the final average. Per the recommendation above, we'll use harmonic mean because we're calculating the speed. So we set up the equation:

$$\frac{4}{\frac{1}{9}+\frac{1}{9}+\frac{1}{9}+\frac{1}{x}} = 10$$

For three laps, Nick's average was 9 mph and we want to solve for x to figure out the speed he needs to maintain to average 10 mph. So we cross multiply and simplify:

$$4 = \frac{30}{9} + \frac{10}{x}$$

Then we collect the constant terms and simplify:

$$4 - \frac{30}{9} = \frac{6}{9} = \frac{2}{3} = \frac{10}{x}$$

Then we cross multiply and simplify again to solve for x.

$$2x = 30 \longrightarrow x = 15$$

So Nick has to run 15 mph to make his goal.

---

[3] From Stanford Math Tournament 2013

Chapter 4

Sequences and Series

First, a few definitions.

DEFINITION 4.1   **Sequence.**

A sequence is any ordered list of objects, either finite or infinite. Examples include $0, 0, 0, \ldots$; $1, 0, 1, 0, \ldots$; $1, 2, 3$; $1, -2, 4, -8, 16, \ldots$; $100, 50, 25, \ldots$; etc.

DEFINITION 4.2   **Series.**

A series is the sum of all terms in any infinite sequence. For example, the sum of the sequence $\left\{\frac{1}{2^n}\right\}$ can be expressed as $\frac{1}{2} + \frac{1}{4} + \frac{1}{8} + \cdots = 1$.

DEFINITION 4.3   **Partial Sum.**

Let $\{a_n\}$ be an infinite sequence of real numbers. The partial sum, $S_N = a_0 + a_1 + a_2 + \cdots + a_N$. The partial sum will always be finite.

An important notation for series is the Sigma notation.

NOTATION 4.1   **Sigma Notation.**

Let $\{a_n\}$ be a infinite sequence of real numbers. Furthermore let m and k be non-negative integers. Then

$$\sum_{i=m}^{k} a_i = a_m + a_{m+1} + a_{m+2} + \cdots + a_{k-1} + a_k.$$

In every Sigma expression there is a subscript expression and a superscript expression. In the subscript of the capital sigma is an *index variable* (i in this example) as well as a *lower bound of summation* or "starting point" (m in this example). In the superscript is an *upper bound of summation* or an "ending point" (k in this example). If there is no upper bound, i.e. the summation is infinite, then the superscript displays ∞.

DEFINITION 4.4   **Converge/Diverge.**

If an infinite series can be expressed as a real number, we say it converges. Otherwise, it diverges.

4.1 Arithmetic Sequences and Series

DEFINITION 4.5   **Arithmetic Sequence.**

An arithmetic sequence is a sequence of the form $a, a+d, a+2d, \ldots$ where the first term is a and the common difference is d.

FORMULA 4.1   **Arithmetic Series.**

The sum of the first n terms of an arithmetic sequence is an arithmetic series equal to

$$\sum_{k=0}^{n-1}(a+kd) = a + (a+d) + (a+2d) + \cdots + (a+(n-1)d)$$
$$= na + \frac{dn(n-1)}{2}.$$

FORMULA 4.2   **A-to-Z Formula.**

The sum in the previous formula can also be expressed in terms of the first term (a) and the last term ($a+(n-1)d$, call it z), such that

$$\sum_{k=0}^{n-1}(a+kd) = \frac{n}{2}(a+z).$$

4.1. ARITHMETIC SEQUENCES AND SERIES

Note: there is no such thing as an infinite arithmetic series, as there exists no such series that converges.

Here's an example of a somewhat difficult problem using arithmetic sequences and series:

EXAMPLE 4.1 An arithmetic progression is formed by five distinct prime numbers. What is the least possible sum of those five numbers?[1]

Solution: The only way to start this problem is by listing off primes in hopes that an arithmetic sequence will arise. We know the correct answer comes from the first 5 term arithmetic sequence starting with he smallest prime number because the question asks for the "least possible sum". The first primes are

$$2, 3, 5, 7, 11, 13, 17, 19, 23, 29, 31, \ldots$$

Already from this we note that $5+6 = 11$, $11+6 = 17$, $17+6 = 23$, and $23 + 6 = 29$. So we have an arithmetic sequence

$$\{5, 11, 17, 23, 29\}$$

which starts at 5 and has a common difference of 6.

We can then use our formula for an arithmetic series to find the sum.

$$na + \frac{dn(n-1)}{2}$$

In this case $n = 5$, the number of terms in the series, $a = 5$ the first term of the series, and $d = 6$ the common difference of the terms. We substitute and solve to find the sum of the arithmetic progression.

$$5 \cdot 5 + \frac{6 \cdot 5 \cdot (5-1)}{2} = 25 + \frac{120}{2} = 25 + 60 = 85$$

So the answer is 85.

---

[1]From MATHCOUNTS 2014 School Compeition, Sprint Round

4.2 Geometric Sequences and Series

DEFINITION 4.6   **Geometric Sequence.**

A geometric sequence is a sequence of the form $a, ar, ar^2, \ldots$ where the first term is a and the common ratio is r.

FORMULA 4.3   **Finite Geometric Series.**

The sum of the first n terms of a geometric sequence is equal to

$$\sum_{k=0}^{n-1} ar^k = a + ar + ar^2 + \cdots + ar^{n-1} = \frac{a(1-r^n)}{1-r}$$

FORMULA 4.4   **Infinite Geometric Series.**

The sum of all the terms of a geometric sequence is equal to

$$\sum_{k=0}^{\infty} ar^k = a + ar + ar^2 + \cdots = \frac{a}{1-r}$$

if and only if $|r| < 1$.

If $|r| \geq 1$, the series will necessarily diverge!

Infinite geometric series can come up in a variety of contest questions, such as economics related questions where a percentage is applied infinitely:

EXAMPLE 4.2   **Some US companies reward their overseas employees by "grossing up" their paychecks by a certain percentage to cover the income tax taken out. Note, however, that when the company increases their paycheck, the amount of taxes they have to cover increases, and the "gross up" needs to be grossed up as well, and so on. If an employee is taxed at a flat rate of 25% of their paycheck, by what percentage does the company need to gross up their pay check to completely cover all income tax?**[2]

Solution: This is clearly a case where a percentage is applied infinitely. Let p be the value of the pay check before taxes and before it

---
[2]TAMU Math Contest EF 2010

4.3. SPECIAL PARTIAL SUMS AND SERIES

has been grossed up. Then the first gross up will result in a pay check with a value of $p+.25p$. But now the $.25p$ has to be grossed up as well, resulting in a pay check with a value of $p+.25p+.25^2p$. This will occur continuously, such that an infinite geometric series results. Using the formula above, the value of the grossed up pay check will be

$$\frac{P}{1-\frac{1}{4}} = \frac{4}{3}P.$$

The question asks "by what percentage does the company need to gross up their pay check...," which translates into a "percent change" question. The percent change is expressed by

$$\frac{\frac{4}{3}P - P}{P} = \frac{1}{3} = 33\frac{1}{3}\%.$$

The company needs to gross up their paycheck by $33\frac{1}{3}\%$.

4.3 Special Partial Sums and Series

FORMULA 4.5 **The Sum of the First n Natural Numbers.**

$$\sum_{k=0}^{n} k = 1 + 2 + 3 + 4 + \cdots + (n-1) + n = \frac{n(n+1)}{2}$$

This formula was first discovered by Gauss in grade school. His teacher would have the class add up the first, say, 100 numbers as punishment. Gauss wrote the sum of the natural numbers in an ascending progression and called it S.

$$S = 1 + 2 + 3 + \cdots + 99 + 100.$$

Then Gauss wrote the same sum S backwards. Finally he added the two alike sums together and found something amazing:

	S	=	1	+	2	+	3	+	$\cdots$	+	99	+	100
+	S	=	100	+	99	+	98	+	$\cdots$	+	2	+	1
	2S	=	101	+	101	+	101	+	$\cdots$	+	101	+	101

Dividing by 2, Gauss found S to be equal to

$$\frac{100\,(101)}{2} = 5050$$

which is in agreement with the formula above.

FORMULA 4.6 **The Sum of the Squares of the First n Natural Numbers.**

$$\sum_{k=1}^{n} k^2 = 1^2 + 2^2 + 3^2 + 4^2 + \cdots + (n-1)^2 + n^2 = \frac{n(n+1)(2n+1)}{6}$$

FORMULA 4.7 **The Sum of the Cubes of the First n Natural Numbers.**

$$\sum_{k=1}^{n} k^3 = 1^3 + 2^3 + 3^3 + 4^3 + \cdots + (n-1)^3 + n^3 = \left(\frac{n(n+1)}{2}\right)^2$$

FORMULA 4.8 **The Sum of the First n Odd Numbers.**

$$\sum_{k=1}^{n} (2k-1) = 1 + 3 + 5 + \cdots + 2n - 1 = n^2$$

COROLLARY 4.9 **The Difference of two Consecutive Squares.**

$$(n)^2 - (n-1)^2 = 2n - 1$$

This can be shown using dots in a square matrix. Starting with one dot, each time the next odd number of dots is added around one corner of the matrix, the next size larger square is created. Consult Figure 4.

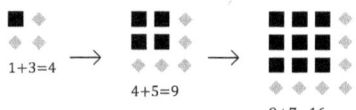

Figure 4

4.4 The Fibonacci Sequence

DEFINITION 4.7 **The Fibonacci Sequence.**

The Fibonacci sequence is a recursively defined sequence which is defined as the sequence $\{f_n\}$, where, $f_1 = 1$, $f_2 = 1$, and when $n \geq 3$

$$f_{n+1} = f_n + f_{n-1}.$$

The first Fibonacci numbers are

$1, 1, 2, 3, 5, 8, 13, 21, 34, 55, 89, 144, 233, 377, 610, 987, 1597, 2584, \ldots$

4.4. THE FIBONACCI SEQUENCE

The Fibonacci numbers have the following basic properties:

- Every 3rd number of the sequence is even and more generally, every kth number of the sequence is a multiple of F_k.

- As n approaches infinity, $\frac{f_{n+1}}{f_n}$ approaches the golden ratio, $\varphi = \frac{1+\sqrt{5}}{2}$. (Not to be confused with Euler's Totient function, $\varphi(n)$.)

- The sum of the first n Fibonacci numbers is the $f_{n+2} - 1$.

- The sum of the squares of the first n Fibonacci numbers is the product $f_n f_{n+1}$.

The Fibonacci numbers can be used in other areas of math as well. For example:

- **Geometry:** Any four consecutive Fibonacci numbers, $f_n, f_{n+1}, f_{n+2}, f_{n+3}$, can form a Pythagorean triple in this way:

$$\begin{align} a &= f_n f_{n+3} \\ b &= 2 f_{n+1} f_{n+2} \\ c &= f_{n+1}^2 + f_{n+2}^2 \ni \\ c^2 &= a^2 + b^2 \end{align}$$

Try it with 1, 2, 3, and 5 to confirm!

- **Combinatorics:** f_{n+2} = number of subsets of $1, 2, ..., n$ that contain no consecutive integers.

- **Combinatorics:** f_{n+1} = number of tilings of a $2 \times n$ rectangle by 2×1 dominoes.

- **Combinatorics:** $f_{n+1} = \binom{f_n}{0} + \binom{f_{n-1}}{1} + \binom{f_{n-2}}{2} + \cdots + \binom{f_{n-k}}{k}$, where k is the greatest integer such that $n - k \geq k$.

Chapter 5
Algebra

5.1 Polynomials

5.1.1 Quadratic Functions

FORMULA 5.1 **The Quadratic Formula.**

Given a quadratic function $ax^2 + bx + c = 0$, the solutions for x are given by the quadratic formula:

$$x = \frac{-b \pm \sqrt{b^2 - 4ac}}{2a}.$$

DEFINITION 5.1 **Discriminant.**

The **discriminant** D of the quadratic equation is $b^2 - 4ac$.

The number and kind of roots are dependent on the value of the discriminant:

Value of Discriminant	Nature of Roots Example
$b^2 - 4ac > 0$	Two distinct real roots $x^2 + 5x + 6 = 0$, $D = 1$, $x = -2$ or $x = -3$
$b^2 - 4ac = 0$	One real double root $x^2 + 4x + 4 = 0$, $D = 0$, $x = -2$
$b^2 - 4ac < 0$	Two conjugate complex roots $x^2 - 2x + 2 = 0$, $D = -4$, $x = 1 \pm i$

FORMULA 5.2   **Product and Sums of Roots of a Quadratic Function.**

If r_1 and r_2 are the roots of $ax^2 + bx + c = 0$, then $r_1 r_2 = \frac{c}{a}$ and $r_1 + r_2 = \frac{-b}{a}$. In the case of a monic, i.e. $a = 1$, we get the product of the roots to be c and the sum of the roots to be $-b$. This generalizes to cubics, etc. See the Sums and Roots of a Cubic Function formula.

FORMULA 5.3   **Maximum or Minimum of a Quadratic Function.**

If $y = f(x) = ax^2 + bx + c = 0$, $a \neq 0$, then the graph of f is a parabola with vertex at $x = -\frac{b}{2a}$. If $a > 0$, then the vertex is the point on the graph of f with the least y value. If $a < 0$, then the vertex is the point in the graph of f with the greatest y value. In either case, the x value at which f attains its minimum or maximum is $x = -\frac{b}{2a}$, while the actual minimum or maximum value of f is $f\left(-\frac{b}{2a}\right)$.[1]

We give some examples from the AMC12:

EXAMPLE 5.1   Both roots of the quadratic equation $x^2 - 63x + k = 0$ are prime numbers. The number of possible values of k is[2]

(A)   0   (B)   1   (C)   2   (D)   4   (E)   more than 4

Solution: In this problem we can use the fact that the two roots sum to be the negative of the coefficient of x (in this case 63) and that the two roots multiply to be the constant of the quadratic equation (in this case k). The question boils down to, "How many pairs of primes sum to 63?" We can notice that this number is odd, meaning that it must be the result of a sum of an even and an odd number. Of course 2 is the only even prime, making the other root 61, and making $k = 122$. Thus, 2 and 61 is the *only* pair that works $\Longrightarrow$(B).

---

[1] Calculus students can easily recognize the nature of this vertex. Since the vertex occurs when $f'(x) = 0$, we differentiate the function to get $2ax + b = 0$ and solve, finding $x = -\frac{b}{2a}$.

[2] From AMC12A 2002

5.1. POLYNOMIALS

EXAMPLE 5.2 The parabola $y = ax^2 + bx + c$ has vertex (p, p) and y-intercept $(0, -p)$, where $p \neq 0$. What is b?[3]

(A) $-p$ (B) 0   (C) 2   (D) 4   (E) p

Solution: Let's start by using the y-intercept which is given to us. So we know that $-p = y = a(0)^2 + b(0) + c = c$, making our parabola now $y = ax^2 + bx - p$. Let's move to the vertex now. We can use the formula for the vertex of a parabola, i.e. $x = -\frac{b}{2a}$. However we are given that $x = p$ at the vertex, so $p = -\frac{b}{2a}$ and $a = -\frac{b}{2p}$. Now we can use the vertex point in the parabola's equation, putting everything in terms of b and p.

$$
\begin{aligned}
y = p &= a(p)^2 + b(p) - p & &[y = p, \text{ plug in } p \text{ for } x] \\
0 &= ap^2 + bp - 2p & &[\text{subtract over the } p] \\
0 &= ap^2 + (b-2)p & &[\text{factor out } p] \\
0 &= \left(-\frac{b}{2p}\right)p^2 + (b-2)p & &[\text{substitute for } a] \\
0 &= \left(-\frac{b}{2}\right)p + (b-2)p & &[\text{cancel the superfluous } p\text{'s}] \\
0 &= \left(-\frac{b}{2}\right) + b - 2 & &[\text{divide away } p, \text{ ok because } p \neq 0] \\
2 &= \frac{b}{2} & &[\text{add 2}, -\frac{b}{2} + b = \frac{b}{2}] \\
4 &= b \Longrightarrow (D) & &[\text{multiply by 2}]
\end{aligned}
$$

5.1.2 Cubic Functions

DEFINITION 5.2 **Cubic Function.**

A cubic function is a polynomial in the form

$$ax^3 + bx^2 + cx + d = 0.$$

There exists a cubic formula; however, memorizing such a formula would be considered an imprudent use of time.

FORMULA 5.4 **Product and Sums of Roots of a Cubic Function.**

---
[3]From AMC12B 2006

For the cubic $ax^3 + bx^2 + cx + d = 0$, the product of the roots is $\frac{-d}{a}$, the sum of the product of all possible combinations of two roots, i.e. $r_1 r_2 + r_2 r_3 + r_1 r_3$, is $\frac{c}{a}$ and the sum of the individual roots is $\frac{-b}{a}$.

FORMULA 5.5 **Relative Maximum and Minimum of a Cubic function.**

If $y = f(x) = ax^3 + bx^2 + cx + d = 0$, $a \neq 0$, then the graph of $f(x)$ may have one "hill" and one "valley." The top of the hill is the relative maximum of the function, and the bottom of the valley is the relative minimum of the function. The x values of the relative maximum and relative minimum of $f(x)$ correspond to the roots of the quadratic function $3ax^2 + 2bx + c = 0$, call them q_1 and q_2, where $q_1 \leq q_2$. If q_1 and q_2 are complex, or if $q_1 = q_2$, there is no relative minimum or relative maximum. Otherwise, there are 2 cases. If $a > 0$ the x value at which f attains its relative maximum is $x = q_1$ and the x value at which f attains its relative minimum is $x = q_2$, making the relative maximum and relative minimum $f(q_1)$ and $f(q_2)$ respectively. Similarly, if $a < 0$ the x value at which f attains its relative minimum is $x = q_1$ and the x value at which f attains its relative maximum is $x = q_2$, making the relative minimum and relative maximum $f(q_1)$ and $f(q_2)$ respectively.

5.1.3 Polynomial Properties

THEOREM 5.1 **The Fundamental Theorem of Algebra.**

Given a polynomial $f(x) = 0$ of degree n, there are exactly n solutions, including multiplicities, for x, each being either real or complex. Note: there will always be an even number of complex roots in a polynomial having only real coefficients.

THEOREM 5.2 **Descartes' Rule of Signs.**

Given a polynomial with real coefficients, with terms arranged in decreasing powers of x, the number of positive real roots is equal to the number of sign changes between

5.1. POLYNOMIALS

consecutive terms of $f(x)$ OR less than the number of sign changes by a multiple of 2.

Similarly, the number of negative real roots is equal to the number of sign changes between consecutive terms of $f(-x)$ OR less than the number of sign changes by a multiple of 2.

The remaining number of roots as given by the Fundamental Theorem of Algebra must be complex.

EXAMPLE 5.3   **What are the possible number of positive real roots, negative real roots, and complex roots of the polynomial function $f(x) = x^5 - 3x^4 + x^3 + 5x^2 + x - 5 = 0$?**

Solution: Using Descartes' Rule of Signs there are sign changes in the following places for $f(x)$

$$f(x) = \underbrace{x^5 - 3x^4}_{+\to-} \overbrace{+ x^3 + 5x^2}^{-\to+} \underbrace{+ x - 5}_{+\to-}$$

That's 3 sign changes, which means that there are either 3 or 1 positive real roots. For the negative roots, apply Descartes' Rule of Signs to $f(-x)$. There are sign changes in the following places for $f(-x)$

$$f(-x) = -x^5 - 3x^4 \overbrace{- x^3 + 5x^2}^{-\to+} \underbrace{}_{+\to-} - x - 5$$

That's 2 sign changes, which means that there are either 2 or 0 negative real roots. We can make a table to summarize the possibilities.

# +ℝ Roots	# −ℝ Roots	# Complex Roots
3	2	0
3	0	2
1	2	2
1	0	4

There are some special factorizations of specific polynomials:

FORMULA 5.6   **Factoring Formulas.**

- **Difference of Squares:** $a^2 - b^2 = (a-b)(a+b)$
- **Square of a Sum:** $a^2 \pm 2ab + b^2 = (a \pm b)^2$

- **Sum of Cubes:** $a^3 + b^3 = (a+b)(a^2 - ab + b^2)$

- **Difference of Cubes:** $a^3 - b^3 = (a-b)(a^2 + ab + b^2)$

EXAMPLE 5.4 Which of the following describes the graph of the equation $(x+y)^2 = x^2 + y^2$?[4]

(A) the empty set (B) one point (C) two lines (D) a circle (E) the entire plane

Solution: Apply the formula of the square of a sum.

$$\begin{aligned} (x+y)^2 &= x^2 + y^2 \\ x^2 + 2xy + y^2 &= x^2 + y^2 \\ 2xy &= 0 \end{aligned}$$

Therefore, either $x = 0$ or $y = 0$. Those equations define two lines (the two axes) $\implies$ (C).

5.2 Exponents and Logarithms

5.2.1 Exponents

FORMULA 5.7 **Laws of Exponents.**

1. $a^n a^m = a^{n+m}$

2. $(a^n)^m = a^{nm}$

3. $(ab)^n = a^n b^n$

4. $\left(\dfrac{a}{b}\right)^n = \dfrac{a^n}{b^n}$

5. $a^{-n} = \dfrac{1}{a^n}$

6. $a^{m/n} = \sqrt[n]{a^m}$

---

[4]From AMC12A 2006

5.2. EXPONENTS AND LOGARITHMS

5.2.2 Logarithms

DEFINITION 5.3   **Logarithm.**

The logarithm, base b, of a number x is equal to the power by which b needs to be raised to get x. In other words, let $b^y = x$, then
$$\log_b x = y.$$

FORMULA 5.8   **Logarithmic Formulas.**

1. $\log_a xy = \log_a x + \log_a y$
2. $\log_a x^n = n \log_a x$
3. $\log_a x/y = \log_a x - \log_a y$
4. **Change of Base:**
$$\log_a x = \frac{\log_b x}{\log_b a}$$

5.2.3 Exponent and Logarithm Examples

Given these rules, it is logical to assume that most contest exponent and logarithm questions are inextricably connected to number theory. Because exponent and logarithm questions are prevalent on tests like the AMC12, we provide some examples here:

EXAMPLE 5.5   **How many positive integers b have the property that $\log_b 729$ is itself a positive integer?**[5]

(A) 0   (B) 1   (C) 2   (D) 3   (E) 4

Solution: Let n be a positive integer. Thus, $\log_b 729 = n$. By definition, $b^n = 729$, and both b and n must be positive integers. The following are ways to express 729 as b^n:

$$729 = 3^6 = 9^3 = 27^2 = 729^1.$$

Therefore, there are 4 (the number of factors of 6) b's that have such a property$\Longrightarrow$ (E).

---
[5]From AMC12 2000.

EXAMPLE 5.6 The solution to the equation $7^{x+7} = 8^x$ can be expressed in the form $x = \log_b 7^7$. What is b?[6]

(A) $\frac{7}{15}$ (B) $\frac{7}{8}$ (C) $\frac{8}{7}$ (D) $\frac{15}{8}$ (E) $\frac{15}{7}$

Solution: In situations where there is an equation with exponential expressions on both sides, it is best to "take the log" of both sides. Note: the log has to be of the same base! For the sake of simplicity use $\log_{10}$ or ln. Therefore:

$$\begin{aligned}
\log(7^{x+7}) &= \log(8^x) & \text{[take log of both sides]} \\
(x+7)\log 7 &= x \log 8 & [\#2] \\
x \log 7 + 7 \log 7 &= x \log 8 & \text{[distribute]} \\
\log 7^7 &= x(\log 8 - \log 7) & [\#2, \text{ subtract over, factor out } x] \\
\frac{\log 7^7}{\log \frac{8}{7}} &= x & [\#3, \text{ divide to isolate } x] \\
\log_{\frac{8}{7}} 7^7 &= x & \text{[change of base } (\#4)]
\end{aligned}$$

Therefore, $b = \frac{8}{7}. \implies$ (C).

EXAMPLE 5.7 Let $f(n) = \log_{2002} n^2$ and let $N = f(11) + f(13) + f(14)$. Which of the following is true about N?[7]

(A) $N < 1$ (B) $N = 1$ (C) $1 < N < 2$
    (D) $N = 2$ (E) $N > 2$

Solution:

$$\begin{aligned}
N &= \log_{2002} 11^2 + \log_{2002} 13^2 + \log_{2002} 14^2 \\
&= \log_{2002}(11^2 \cdot 13^2 \cdot 14^2) & [\#1] \\
2002^N &= (11 \cdot 13 \cdot 14)^2 & \text{[Definition of Logarithm]}
\end{aligned}$$

Since $2002 = (11 \cdot 13 \cdot 14)$, $N = 2 \implies$ (D).

---

[6] From AMC12A 2010.
[7] From AMC12A 2002.

Chapter 6

Geometry

In this chapter, the sections will be separated so that each one has a two-page spread to facilitate comparison between the algebraic formulas about geometrical figures and the visual representation of such formulas. Therefore, this page is intentionally left blank.

6.1 Circles

For the following circle formulas and theorems, let C be a circle with radius r and let θ be a central angle measured in radians.

FORMULA 6.1 **Arclength Formulas.**

- $2\pi r$ = circumference of C (Fig. 5A)

- θr = length of circular arc with central angle θ measured in radians (Fig. 5A)

FORMULA 6.2 **Area Formulas.**

- πr^2 = area of C (Fig. 5B)

- $\theta r^2/2$ = area of a sector of C with central angle θ measured in radians (Fig. 5B)

THEOREM 6.1 **Basic Theorems.**

- Tangents to circle C meet the radius at a right angle. (Fig. 6A)

- Given a point P outside of circle C and points A and B such that $\overline{PB}$ and $\overline{PB}$ are tangential to C, $PA = PB$. (Fig. 6A)

- Any right triangle inscribed in circle C must be such that the hypotenuse is a diameter. (Fig. 6B)

- The angle measure of $\angle ADB$ inscribed in circle C (so $\overline{DA}$ and $\overline{DB}$ are chords of the circle) is half the central angle $\angle AOD$ where O is the center of the circle. (Fig. 6A)

THEOREM 6.2 **Special Features.**

Ptolemy's Theorem

Let $ABCD$ be a quadrilateral which can be inscribed in a circle (a cyclic quadrilateral). Assume the vertices are labeled in clockwise or counterclockwise order. Then

$$AB \cdot CD + AD \cdot BC = AC \cdot BD$$

For another property of cyclic quadrilaterals, see Brahmagupta's Formula.

6.1. CIRCLES

Power of a Point Theorem

Fix a circle C and a point P. The point P may be in, on, or outside of the circle C. Draw **any** line through P. Label the line's points of intersection with the circle by X and Y (for a tangent line, $X = Y$). Then the quantity $PX \cdot PY$ is the same no matter which line was chosen. In Figure 6B, for instance, $PX \cdot PY = PW \cdot PZ = PT^2$.

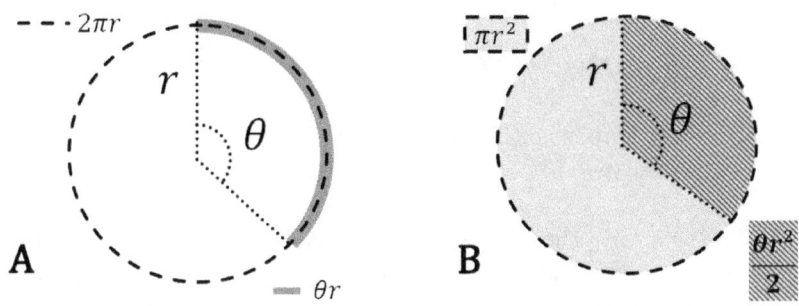

Figure 5

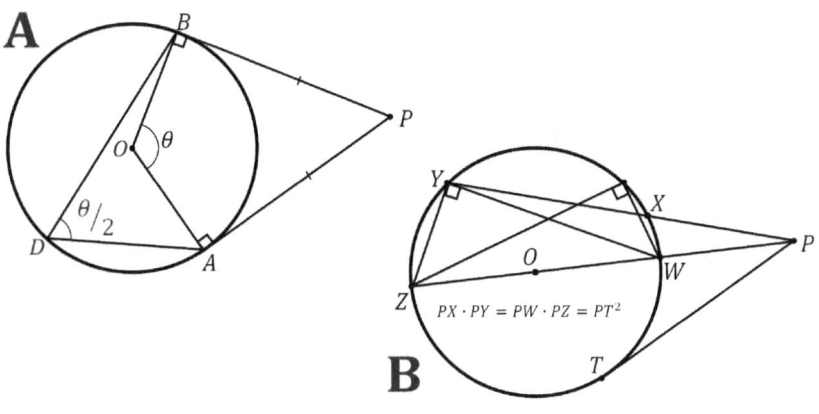

Figure 6

6.2 Quadrilaterals

In this section, let Q be a quadrilateral. Notationally: $b =$ base of the parallelogram, $h =$ height of parallelogram, $d_Q =$ a diagonal of a quadrilateral, and $s =$ side length of a square.

FORMULA 6.3   **Area Formulas.**

- If Q is a rectangle or parallelogram, the area of $Q = bh$. (Fig. 7A, 7C)

- If Q is a square, the area of $Q = s^2$. (Fig. 7B)

- If Q is a trapezoid, the area of $Q = \frac{(b_1+b_2)h}{2}$ (the average of the bases multiplied by the height). (Fig. 7E)

- If Q is a kite, rhombus, or any quadrilateral where the two diagonals are perpendicular, the area of $Q = \frac{(d_1 d_2)}{2}$ (half the product of the diagonals). (Fig. 7D)

- **Brahmagupta's Formula:** If Q is a cyclic quadrilateral, that is, if Q can be circumscribed such that each vertex of Q lies on the circle, the area of $Q = \sqrt{(S-a)(S-b)(S-c)(S-d)}$ where a, b, c, and d are the side lengths of Q, and S is the semiperimeter ($\frac{a+b+c+d}{2}$) of Q.

FORMULA 6.4   **Diagonal Formulas and Theorems.**

- If Q is a rectangle, $d_Q = \sqrt{b^2 + h^2}$. (Fig. 7A)

- If Q is a square, $d_Q = s\sqrt{2}$. (Fig. 7B)

- If Q is a square or rectangle, the two diagonals of Q are congruent. (Fig. 7A, 7B)

- If Q is a square, rectangle, parallelogram, kite, or rhombus, the two diagonals of Q bisect each other. (Fig. 7A, 7B, 7C, 7D)

6.2. QUADRILATERALS

A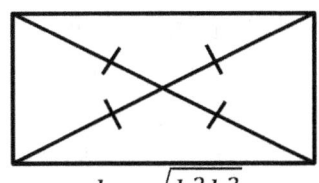

B

C

D

E

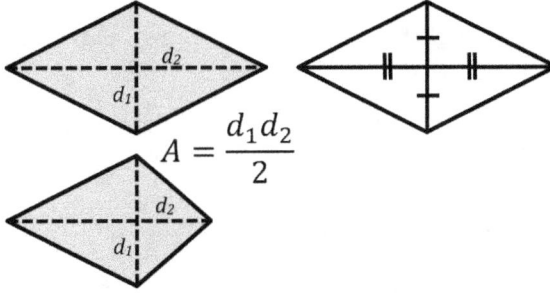

Figure 7

6.3 Triangles

In this section, let $\triangle ABC$ be a triangle, $b = $ the base of that triangle (unless b is a side), $h = $ the height of that triangle, $s = $ side length of an equilateral triangle.

FORMULA 6.5 **Area Formulas.**

- $\dfrac{ab \sin C}{2} = $ area of $\triangle ABC$ with sides a and b which meet at angle C. (Fig. 8A)

- $\dfrac{bh}{2} = $ area of any $\triangle ABC$. (Fig. 8B)

- $\dfrac{s^2 \sqrt{3}}{4} = $ area of an equilateral triangle of side s. (Fig. 8B)

- **Heron's Formula:** $\sqrt{S(S-a)(S-b)(S-c)} = $ area of $\triangle ABC$ with sides a, b, and c where $S = \frac{a+b+c}{2} = $ the semiperimeter of the triangle.

THEOREM 6.3 **Pythagorean Theorem and its Corollaries.**

- A triangle with sides $a \leq b \leq c$ is a **right triangle** if and only if $a^2 + b^2 = c^2$. (Fig. 8C)

- A triangle with sides $a \leq b \leq c$ is an **acute triangle** if and only if $a^2 + b^2 > c^2$. (Fig. 8C)

- A triangle with sides $a \leq b \leq c$ is an **obtuse triangle** if and only if $a^2 + b^2 < c^2$. (Fig. 8C)

FORMULA 6.6 **Special Right Triangles.**

- **Isosceles Right Triangle:** In an isosceles right triangle, the two non-right angles measure $45°$, the two legs have equal length x, and the hypotenuse has $\sqrt{2}$ times that length: $x\sqrt{2}$. (Fig. 8D)

- **30–60–90 Right Triangle:** In a right triangle where one of the non-right angles measures $30°$ and the other measures $60°$, if the short leg measures x, the hypotenuse will measure $2x$ and the other leg will measure $x\sqrt{3}$. (Fig. 8D)

6.3. TRIANGLES

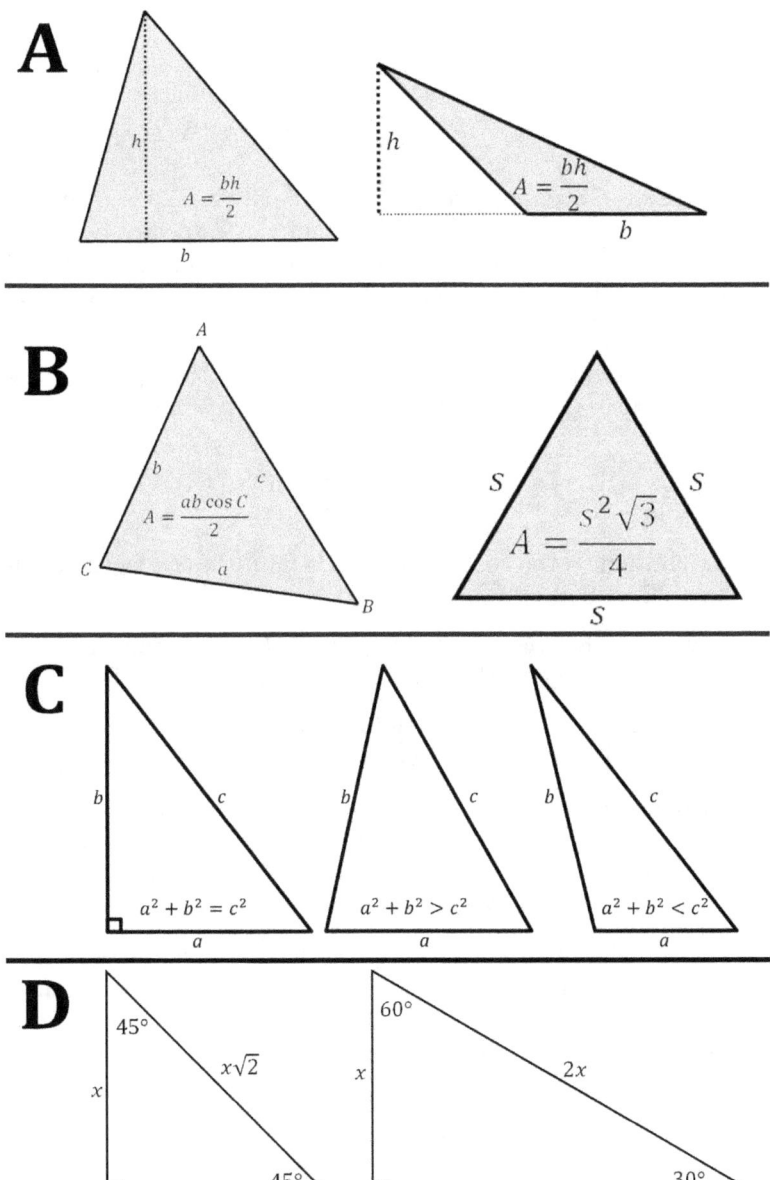

Figure 8

THEOREM 6.4   **Special Features.**

- In $\triangle ABC$ let $\overrightarrow{CD}$ bisect $\angle BCA$ where D lies on segment $\overline{AB}$. (Fig. 9A) Then
$$\frac{DB}{BC} = \frac{DA}{AC}$$

- **Median Concurrence Theorem:** Given $\triangle ABC$, and midpoints $D, E,$ and F on $\overline{AB}, \overline{BC}$ and $\overline{CA}$ respectively, the **medians** $\overline{AE}, \overline{BF},$ and $\overline{CD}$ all intersect at a single point P, called the centroid of the triangle, such that $AP = 2PE, BP = 2PF,$ and $CP = 2PD$. (Fig. 9B)

- Given a right triangle, $\triangle ABC$, where C is the right angle, and the altitude $\overline{CQ}$ to the hypotenuse, the following are true
$$CQ^2 = AQ \cdot BQ$$
$$CB^2 = AB \cdot BQ$$
$$CA^2 = AB \cdot AQ$$
in addition to the superficial results from the Pythagorean theorem. (Fig. 9C)

- Given a right triangle, $\triangle ABC$ where C is the right angle, and the median $\overline{CM}$ to the hypotenuse, $CM = 2AB$.[1] (Fig. 9D)

- Given $\triangle ABC$, and midpoints D and E on $\overline{AB}$ and $\overline{AC}$ respectively, $\overline{DE}$ is parallel to $\overline{BC}$ and $DE = \frac{BC}{2}$. (Fig. 9E)

- **Stewart's Theorem:** In $\triangle ABC$, let D be a point on $\overline{BC}$, making $\overline{AD}$ a cevian. Call $AD = d$, $BD = m$, and $CD = n$. (Fig. 9F) Then,
$$b^2 m + c^2 n = a(d^2 + mn).$$

- **Ceva's Theorem:** A line segment joining a vertex of a triangle to any point on the opposite side is called a **cevian**. (So a median is a special kind of cevian.) In $\triangle ABC$ let three cevians $AD, BE,$ and CF be given. Then it follows that if and only if the cevians are concurrent
$$\frac{BD}{DC} \cdot \frac{CE}{EA} \cdot \frac{AF}{FB} = 1.$$
Recall that concurrent means that when extended as lines, they all intersect in one same point. (Fig. 9G)

---

[1]This is related to the fact that a circumscribed right triangle has a hypotenuse which is a diameter.

6.3. TRIANGLES

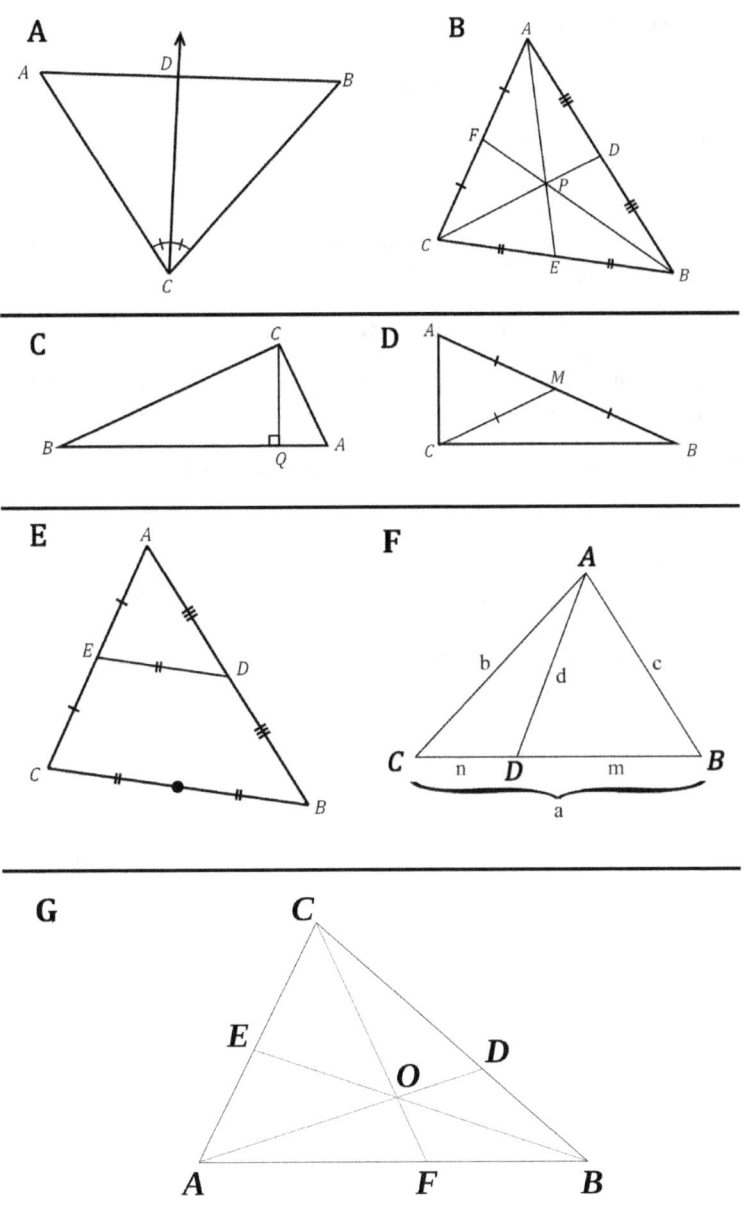

Figure 9

6.4 Angles

THEOREM 6.5 **Planar Angles.**

- Opposite angles of two intersecting angles are equal. (Fig. 13A)
- Parallel lines intersect a given line in equal angles. (Fig. 13A)
- An exterior angle of a triangle equals the sum of the opposite interior angles. (Fig. 13B)
- The sum of the angles of a triangle equals 180°.
- The sum of the interior angles of an n side polygon equals (in degrees) $180(n-2)$.
- Each interior angle of a regular polygon with n sides measures (in degrees) $\frac{180(n-2)}{n}$.

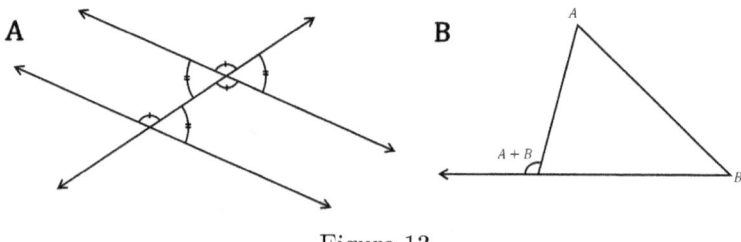

Figure 13

6.5 Three Dimensional Solids

FORMULA 6.7 **Surface Area Formulas.**

- $4\pi r^2 =$ surface area of a sphere of radius r
- $2\pi r(r+h) =$ surface area of a right cylinder of radius r and height h
- $\pi r s =$ lateral surface area of a right circular cone of radius r and slantheight s. (Fig. 14)
- Note: by the Pythagorean Theorem, $s^2 = \sqrt{h^2 + r^2}$.

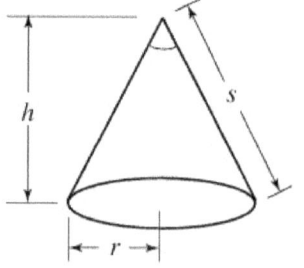

Figure 14

6.5. THREE DIMENSIONAL SOLIDS

FORMULA 6.8   **Volume Formulas.**

- $\dfrac{4\pi r^3}{3}$ = volume of a sphere of radius r
- s^3 = volume of a cube of side length s
- Bh = volume of a cylinder with base area B and height h
- $\dfrac{Bh}{3}$ = volume of a pyramid or cone with base area B and height h

FORMULA 6.9   **The Diagonal of a Rectangular Prism.**

Given a rectangular prism with edge lengths a, b, and c, the diagonal (or the diameter of the circumscribing sphere) that goes through the center of the rectangular prism has length
$$\sqrt{a^2 + b^2 + c^2}.$$
This is the result of the double application of the Pythagorean theorem.

DEFINITION 6.1   **Platonic Solid.**

A platonic solid is a convex polyhedron in which every face is a congruent regular polygon, the same number of faces meet at every vertex, every edge is congruent, and every angle between every edge is congruent.

There are only five Platonic Solids. They are represented in Figure 15 below.

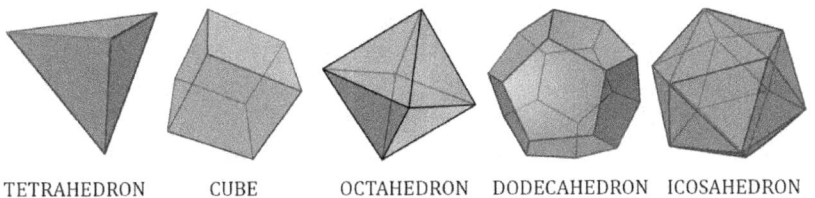

TETRAHEDRON   CUBE   OCTAHEDRON   DODECAHEDRON   ICOSAHEDRON

Figure 15—The Platonic Solids

FORMULA 6.10 Euler's Formula for Polyhedra.

Given a convex polyhedron with V vertices, E edges, and F faces,
$$V - E + F = 2.$$

In lieu of a formal proof, let's look at the Platonic Solids to verify the legitimacy of Euler's Formula.

Polyhedron	V	E	F	$V - E + F$
Tetrahedron	4	6	4	2
Cube	8	12	6	2
Octahedron	6	12	8	2
Dodecahedron	20	30	12	2
Icosahedron	12	30	20	2

This strong evidence gives credence to Euler's Formula.

6.6 Geometry Examples

The following are all questions from previous AMC12 Exams.

EXAMPLE 6.1 Each of the small circles in the figure has radius one. The innermost circle is tangent to the six circles that surround it, and each of those circles is tangent to the large circle and to its small-circle neighbors. Find the area of the shaded region.[2]

(A) π **(B)** 1.5π **(C)** 2π **(D)** 3π **(E)** 3.5π

Solution: Because the circles are all tangent to each other, and the outside circles are tangent to the big circle, three radii of the small circle are equal to the radius of the big circle, i.e. the radius of the big circle is 3. The area of the shaded region is equal to the difference between the area of the big circle and the area of the 7 smaller circles. Therefore, the area of the shaded region is $9\pi - 7\pi = 2\pi \implies$ (C).

---
[2]From AMC12A 2002

6.6. GEOMETRY EXAMPLES

EXAMPLE 6.2 An annulus is the region between two concentric circles. The concentric circles in the figure have radii b and c, with $b > c$. Let OX be a radius of the larger circle, let XZ be tangent to the smaller circle at Z, and let OY be the radius of the larger circle that contains Z. Let $a = XZ$, $d = YZ$, and $e = XY$. What is the area of the annulus?[3]

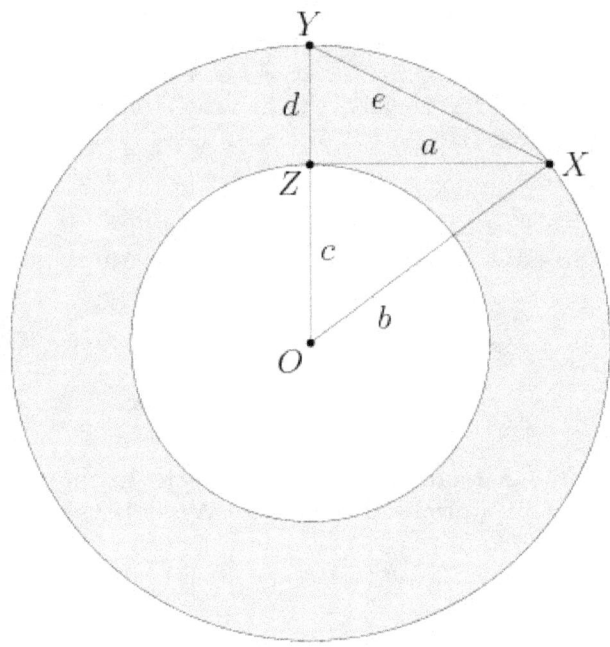

(A) πa^2 (B) πb^2 (C) πc^2 (D) πd^2 (E) πe^2

Solution: The area of the larger circle is πb^2, and the area of the smaller circle is πc^2. Therefore, the area of the annulus is $\pi(b^2 - c^2)$. But because $\overline{XZ}$ is tangent to the smaller circle, $\triangle OXZ$ is right. By the Pythagorean theorem, $b^2 - c^2 = a^2$, therefore the area is also equal to $\pi a^2 \implies$ (A).

---
[3] From AMC12B 2004

EXAMPLE 6.3 Let v, w, x, y, and z be the degree measures of the five angles of a pentagon. Suppose that $v < w < x < y < z$ and v, w, x, y, and z form an arithmetic sequence. Find the value of x.[4]
(A) 72   (B) 84   (C) 90   (D) 108   (E) 120

Solution: Because it's in the middle, let's work with x and put all the other variables in terms of x. Because the five angles form an arithmetic sequence, the difference from one to the next is a constant, call it k. Therefore, $v = x - 2k, w = x - k, x = x, y = x + k$, and $z = x + 2k$. Use the formula above to find the total number of degrees in a pentagon: $(5-2)180 = 540$. Solve for x:

$$\begin{aligned} v + w + x + y + z &= 540 \\ x - 2k + x - k + x + x + k + x + 2k &= 540 \\ 5x &= 540 \\ x &= 108 \Longrightarrow \text{(D)}. \end{aligned}$$

EXAMPLE 6.4 A triangle with side lengths in the ratio $3:4:5$ is inscribed in a circle of radius 3. What is the area of the triangle?[5]
(A) 8.64   (B) 12   (C) 5π (D) 17.28   (E) 18

Solution: Because $3^2 + 4^2 = 5^2$, we know that the triangle is right. Since it's inscribed in a circle, the triangle's hypotenuse is equal to the diameter of the circle, which in this case is 6. The other two legs are in proportion, i.e. $\frac{6}{5}(4) = \frac{24}{5}$ and $\frac{6}{5}(3) = \frac{18}{5}$. Using the formula for the area of a triangle, the area is equal to $\frac{\frac{24}{5} \cdot \frac{18}{5}}{2} = \frac{216}{25}$ and

$$8 < \frac{216}{25} < 9 \Longrightarrow (A).$$

---

[4]From AMC12B 2002
[5]From AMC12A 2007

6.6. GEOMETRY EXAMPLES

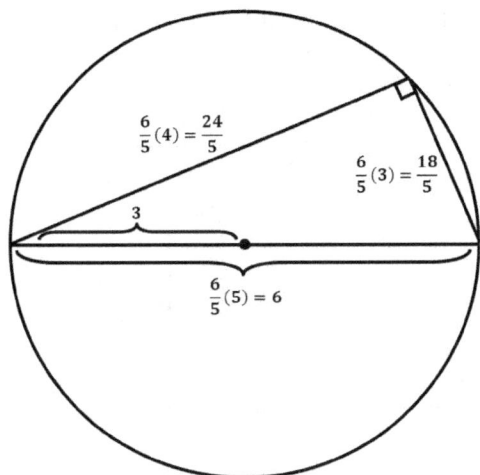

Solution to Example 6.4

EXAMPLE 6.5 An ice cream cone consists of a sphere of vanilla ice cream and a right circular cone that has the same diameter as the sphere. If the ice cream melts, it will exactly fill the cone. Assume that the melted ice cream occupies 75% of the volume of the frozen ice cream. What is the ratio of the cone's height to its radius?[6]
(A) $2:1$ (B) $3:1$ (C) $4:1$ (D) $16:3$ (E) $6:1$

Solution: Let h be the height of the cone and r be the radius of the cone and the ice cream sphere. The volume of the cone, therefore, is $\frac{1}{3}\pi r^2 h$, and the volume of the sphere is $\frac{4}{3}\pi r^3$. Frozen, the ice cream has the volume $\frac{4}{3}\pi r^3$, but when it melts, it's only 75% of $\frac{4}{3}\pi r^3$. It's also $\frac{1}{3}\pi r^2 h$ when it melts. We can set the two equations equal to each other:

$$\frac{3}{4} \cdot \frac{4}{3}\pi r^3 = \frac{1}{3}\pi r^2 h$$
$$\pi r^3 = \frac{1}{3}\pi r^2 h$$
$$3 = \frac{\pi r^2 h}{\pi r^3}$$
$$3 = \frac{h}{r} \implies (B)$$

---
[6]From AMC12B 2003

EXAMPLE 6.6   Rhombus $ABCD$ is similar to rhombus $BFDE$. The area of rhombus $ABCD$ is 24, and $\angle BAD = 60°$. What is the area of rhombus $BFDE$?[7]

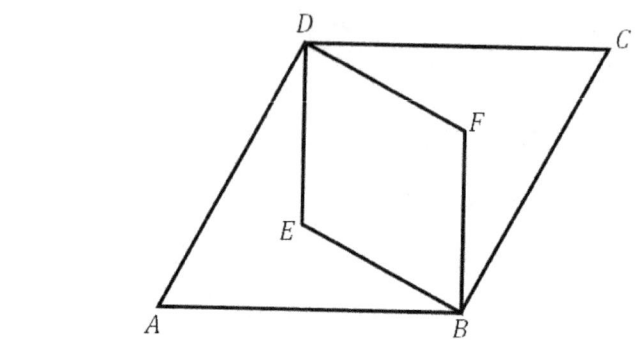

(A) 6   (B) $4\sqrt{3}$ (C) 8   (D) 9   (E) $6\sqrt{3}$

Solution: Note: there are many different ways to solve this problem. We'll start by using the formula for the area of a rhombus: $24 = \frac{DB \cdot AC}{2}$. Now, because of the definition of a rhombus and the fact that $\angle BAD = 60°$, each rhombus can be thought of as two equilateral triangles.

In $ABCD$ $AB = BC = CD = DA = BD$. The midpoint of $\overline{BD}$ to A is the altitude of the equilateral triangle $\triangle ABD$, which has length $\frac{DB\sqrt{3}}{2}$.[8] Therefore, $AC = DB\sqrt{3}$.

Now, because the two rhombi are similar, $\frac{AC}{DB} = \frac{DB}{EF}$, making

$$EF = \frac{DB^2}{DB\sqrt{3}},$$

and the area of the small rhombus

$$\frac{DB \cdot \frac{DB^2}{DB\sqrt{3}}}{2} = \frac{DB^2}{2\sqrt{3}}.$$

Let's solve for DB^2 using the first equation for the area of the $ABCD$: $24 = \frac{DB \cdot DB\sqrt{3}}{2} \iff DB^2 = \frac{48}{\sqrt{3}}$. Substitute in the formula for the area of the small rhombus to get $\frac{48}{2 \cdot 3} = 8 \implies$ (C).

---

[7]From AMC12B 2006

[8]This is true because inside each equilateral triangle there are two 30–60–90 right triangles. The long leg corresponds to the altitude.

Chapter 7

Trigonometry

7.1 Definitions

DEFINITION 7.1   **The Fundamentals of Trigonometry.**

In trigonometry, we talk about angles in a right triangle. Let θ be the non-right angle $\angle A$ in the right triangle $\triangle ABC$, where B is the right angle. Then we call the opposite leg BC and the adjacent leg AB, and of course the hypotenuse is AC. (Fig. 16) We give the following definitions:

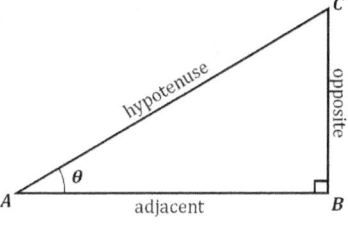

Figure 16

- **Cosine:**
$$\cos \theta = \frac{\text{adjacent}}{\text{hypotenuse}} = \frac{AB}{AC}$$

- **Sine:**
$$\sin \theta = \frac{\text{opposite}}{\text{hypotenuse}} = \frac{BC}{AC}$$

- **Tangent:**
$$\tan \theta = \frac{\text{opposite}}{\text{adjacent}} = \frac{BC}{AB} = \frac{\sin \theta}{\cos \theta}$$

- **Secant:**
$$\sec \theta = \frac{1}{\cos \theta}$$

- **Cosecant:**
$$\csc \theta = \frac{1}{\sin \theta}$$

- **Cotangent:**
$$\cot \theta = \frac{1}{\tan \theta}$$

If you have a hard time remembering the three basic functions, just remember **SOHCAHTOA**. **S**in equals **O**pposite over **H**ypotenuse, **C**os equals **A**djacent over **H**ypotenuse, and **T**an equals **O**pposite over **A**djacent.

7.2 Basic Values and Identities

FORMULA 7.1   **Special Trigonometric Values.**

For certain θ, values for $\cos\theta$, $\sin\theta$, and $\tan\theta$ should be memorized.

θ	$\cos\theta$	$\sin\theta$	$\tan\theta$
$0°$	1	0	0
$30°$	$\frac{\sqrt{3}}{2}$	$\frac{1}{2}$	$\frac{1}{\sqrt{3}}$
$45°$	$\frac{\sqrt{2}}{2}$	$\frac{\sqrt{2}}{2}$	1
$60°$	$\frac{1}{2}$	$\frac{\sqrt{3}}{2}$	$\sqrt{3}$
$90°$	0	1	—
$120°$	$-\frac{1}{2}$	$\frac{\sqrt{3}}{2}$	$-\sqrt{3}$
$135°$	$-\frac{\sqrt{2}}{2}$	$\frac{\sqrt{2}}{2}$	-1
$150°$	$-\frac{\sqrt{3}}{2}$	$\frac{1}{2}$	$-\frac{1}{\sqrt{3}}$
$180°$	-1	0	0

θ	$\cos\theta$	$\sin\theta$	$\tan\theta$
$180°$	-1	0	0
$210°$	$-\frac{\sqrt{3}}{2}$	$-\frac{1}{2}$	$\frac{1}{\sqrt{3}}$
$225°$	$-\frac{\sqrt{2}}{2}$	$-\frac{\sqrt{2}}{2}$	1
$240°$	$-\frac{1}{2}$	$-\frac{\sqrt{3}}{2}$	$\sqrt{3}$
$270°$	0	-1	—
$300°$	$\frac{1}{2}$	$-\frac{\sqrt{3}}{2}$	$-\sqrt{3}$
$315°$	$\frac{\sqrt{2}}{2}$	$-\frac{\sqrt{2}}{2}$	-1
$330°$	$\frac{\sqrt{3}}{2}$	$-\frac{1}{2}$	$-\frac{1}{\sqrt{3}}$
$360°$	1	0	0

FORMULA 7.2   **Basic Trigonometric Identities.**

- $\sin^2\theta + \cos^2\theta = 1$
- $1 + \tan^2\theta = \sec^2\theta$
- $1 + \cot^2\theta = \csc^2\theta$
- **Sine of a Sum:**

$$\sin(\theta \pm \psi) = \sin\theta\cos\psi \pm \cos\theta\sin\psi$$

- **Cosine of a Sum:**

$$\cos(\theta \pm \psi) = \cos\theta\cos\psi \mp \sin\theta\sin\psi$$

7.3. LAW OF SINES AND LAW OF COSINES

- Tangent of a Sum:

$$\tan(\theta \pm \psi) = \frac{\tan\theta \pm \tan\psi}{1 \mp \tan\theta \tan\psi}$$

- Double angle formula for sine:

$$\sin(2\theta) = 2\sin\theta\cos\theta$$

- Double angle formula for cosine:

$$\begin{aligned}\cos(2\theta) &= \cos^2\theta - \sin^2\theta \\ &= 1 - 2\sin^2\theta \\ &= 2\cos^2\theta - 1\end{aligned}$$

- Half angle formula for sine:

$$\sin^2(\theta/2) = \frac{1 - \cos\theta}{2}$$

- Half angle formula for cosine:

$$\cos^2(\theta/2) = \frac{1 + \cos\theta}{2}$$

7.3 Law of Sines and Law of Cosines

FORMULA 7.3
The Law of Sines.
In triangle ABC (Fig. 17),

$$\frac{\sin A}{a} = \frac{\sin B}{b} = \frac{\sin C}{c}$$

FORMULA 7.4
The Law of Cosines.
In triangle ABC (Fig. 17),

$$c^2 = a^2 + b^2 - 2ab\cos C$$

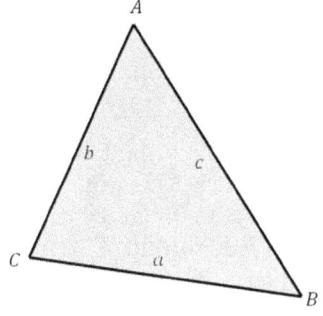

Figure 17

7.4 Examples

We give a few examples:

EXAMPLE 7.1   Suppose $\cos x = 0$ and $\cos(x+z) = 1/2$. What is the smallest possible positive value of z?[1]
(A) 30°   (B) 60°   (C) 90°   (D) 150°   (E) 210°

Solution: Since $\cos x = 0$, $x = 90°$ or $x = 270°$. If $x = 90°$, then the smallest possible value of z that makes $\cos(x+z) = 1/2$ is 210°, making $x + z = 300$. If $x = 270°$, then the smallest possible value of z that makes $\cos(x+z) = 1/2$ is 30°, again making $x + z = 300$. Therefore, the smallest possible value overall, given that we don't know if $x = 90°$ or $x = 270°$ is 30° $\Longrightarrow$ (A).

EXAMPLE 7.2   Points K, L, M, and N lie in the plane of the square $ABCD$ so that AKB, BLC, CMD, and DNA are equilateral triangles. The area of $ABCD$ is 16. What is the area of $KLMN$?[2]

(A)   32

(B) $16 + 16\sqrt{3}$

(C)   48

(D) $32 + 16\sqrt{3}$

(E)   64

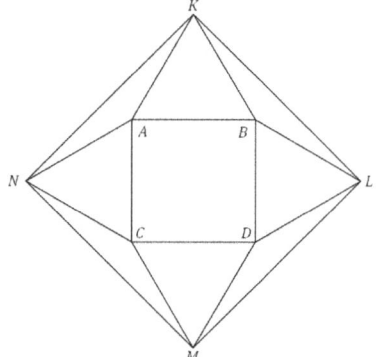

Solution: Note: there are many ways to solve this problem. This solution features the law of cosines. Because the area of $ABCD$ is 16, each side of $ABCD$ is 4. And because each of the triangles adjacent to $ABCD$ is an equilateral triangle, each of their sides is also 4. We can use this fact to find KN in the triangle AKN, and because $KLMN$ is a square, it's area is equal to NK^2. The law of cosines says that

---
[1] From AMC12A 2006. Answers have been converted from radians to degrees.
[2] From AMC12A 2003

7.4. EXAMPLES

$NK^2 = AK^2 + AN^2 - 2AN \cdot AK \cos \angle NAK$. As we noted above, $AK = AN = 4$. Around point A, there are $360°$. $90°$ are the taken by square, and $2 \cdot 60°$ are taken by the two equilateral triangles. That leaves $150°$ for $\angle NAK$. Substitute all these values in our law of cosines expression.

$$\begin{aligned} NK^2 &= 4^2 + 4^2 - 2 \cdot 4 \cdot 4 \cos 150° \\ &= 32 - 32\left(-\frac{\sqrt{3}}{2}\right) = 32 + 16\sqrt{3} \implies (D) \end{aligned}$$

Chapter 8

The Coordinate Plane

8.1 Lines

FORMULA 8.1 **The Slope of a Line.**

Given a line ℓ through two points (x_1, y_1) and (x_2, y_2), the slope m of line ℓ is

$$m = \frac{y_1 - y_2}{x_1 - x_2}.$$

FORMULA 8.2 **Slope-Intercept Form of a Line.**

Given a line ℓ with slope m and y-intercept b, an equation for line ℓ is

$$y = mx + b.$$

FORMULA 8.3 **Point-Slope Form of a Line.**

Given a line ℓ with slope m and a point (x_1, y_1) on the line, an equation for line ℓ is

$$(y - y_1) = m(x - x_1).$$

FORMULA 8.4 **Double Intercept Form of a Line.**

Given a line ℓ with x-intercept a and y-intercept b, an equation for line ℓ is

$$\frac{x}{a} + \frac{y}{b} = 1.$$

FORMULA 8.5 **Perpendicular Lines.**

Two lines with slopes m_1 and m_2 respectively are perpendicular exactly when
$$m_1 \cdot m_2 = -1.$$

FORMULA 8.6 **Distance to a Line.**

The distance d from the point (x, y) to the line $ax + by = c$ is given by
$$d = \frac{|ax + by - c|}{\sqrt{a^2 + b^2}}.$$

8.2 Parabolas

FORMULA 8.7 **Equation of a Parabola.**

Though any quadratic will give the equation of a parabola, the equation
$$y = a(x - h)^2 + k$$
gives the equation of a parabola with vertex (h, k).

FORMULA 8.8 **Vertex of a Parabola.**

The vertex of the parabola $y = ax^2 + bx + c$ occurs when
$$x = \frac{-b}{2a}.$$

8.3 Polygon Tricks

We provide two tricks to find the area of a polygon in the coordinate plane.

FORMULA 8.9 **Pick's Theorem.**

Suppose a polygon $\mathcal{P}$ has all of its vertices given by integer coordinates (a point (a, b) with both a and b integers; such points are called **lattice points**). Then, if n lattice points lie on the polygon itself and m lattice points are contained strictly inside of the polygon, its area is given by the formula:
$$\text{area of } \mathcal{P} = \frac{n}{2} + m - 1.$$

8.3. POLYGON TRICKS

EXAMPLE 8.1 Use Pick's Theorem to find the area of the polygon outlined below.

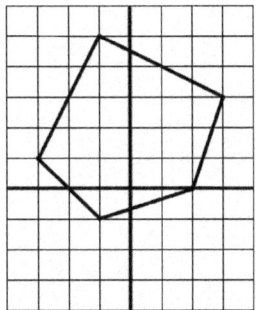

Solution: Since there are 8 lattice points on the polygon and 19 interior lattice points, the area is: $\frac{8}{2} + 19 - 1 = 22$.

FORMULA 8.10 **The Shoestring Formula.**

To find the area of a n-gon with vertices at coordinates $(x_1, y_1), (x_2, y_2), (x_3, y_3), \ldots, (x_n, y_n)$ first create a "shoe" in the following manner:

$$\begin{array}{cccccc} x_1 & x_2 & x_3 & \cdots & x_n & x_1 \\ y_1 & y_2 & y_3 & \cdots & y_n & y_1 \end{array}$$

Then draw diagonals from x_1 to y_2, from x_2 to y_3, etc., and from y_1 to x_2, from y_2 to x_3, etc. Multiply the two coordinates in the forward slashing "shoestrings," and add all the products to get a number F. Similarly, multiply the two coordinates in the backward slashing "shoestrings," and add all those products to get a number B. The area of the polygon will then be $\frac{|F - B|}{2}$.

EXAMPLE 8.2 Find the area of the area of the polygon with vertices at $(3, 0), (4, 2), (3, 6),$ and $(-2, 2)$.
Solution: Start by creating the "shoe" and draw in the "shoestrings."

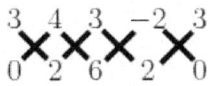

Then $F = 3\cdot 2 + 4\cdot 6 + 3\cdot 2 + (-2)\cdot 0 = 6 + 24 + 6 + 0 = 36$ and $B = 4\cdot 0 + 3\cdot 2 + (-2)\cdot 6 + 3\cdot 2 = 0 + 6 - 12 + 6 = 0$. Thus, using the shoestring formula, the area is $\dfrac{|36-0|}{2} = 18$.

8.4 Miscellaneous

FORMULA 8.11   **Equation of a Circle.**

The equation
$$(x-h)^2 + (y-k)^2 = r^2$$
gives a circle with center (h, k) and radius r in the coordinate plane.

FORMULA 8.12   **Rotation of a Point.**

To rotate a point (x, y) by an angle θ in the counter-clockwise direction take the matrix product

$$\begin{pmatrix} \cos\theta & -\sin\theta \\ \sin\theta & \cos\theta \end{pmatrix} \begin{pmatrix} x \\ y \end{pmatrix}$$

Chapter 9

Problem Solving Strategies

9.1 Number Theory Tips

TIP 1 **Memorize the prime factorization of the current year.** Almost every math competition has at least one question that relates to the year it was written. Most these will invariably involve the number's factors, etc. Memorize the current year's prime factorization so that when that question involving 2015 comes up, you'll be ready (see Example 5.6, for example). Here are the prime factorizations for the next 25 years:

- $2014 = 2 \cdot 19 \cdot 53$
- $2015 = 5 \cdot 13 \cdot 31$
- $2016 = 2^5 \cdot 3^2 \cdot 7$
- $2017 = 2017$
- $2018 = 2 \cdot 1009$
- $2019 = 3 \cdot 673$
- $2020 = 2^2 \cdot 5 \cdot 101$
- $2021 = 43 \cdot 47$
- $2022 = 2 \cdot 3 \cdot 337$
- $2023 = 7 \cdot 17^2$
- $2024 = 2^3 \cdot 11 \cdot 23$
- $2025 = 3^4 \cdot 5^2$
- $2026 = 2 \cdot 1013$
- $2027 = 2027$
- $2028 = 2^2 \cdot 3 \cdot 13^2$
- $2029 = 2029$
- $2030 = 2 \cdot 5 \cdot 7 \cdot 29$
- $2031 = 3 \cdot 677$
- $2032 = 2^4 \cdot 127$
- $2033 = 19 \cdot 107$
- $2034 = 2 \cdot 3^2 \cdot 113$
- $2035 = 5 \cdot 11 \cdot 37$
- $2036 = 2^2 \cdot 509$
- $2037 = 3 \cdot 7 \cdot 97$
- $2038 = 2 \cdot 1019$
- $2039 = 2039$

9.2 Combinatorics Tips

TIP 2 **Use the hangman method.** The hangman method really refers to the general process of drawing a dash to indicate a space for selection or a choice. If you look in the combinatorics section, most of the graphics include a space to indicate a choice or a selection. Using the hangman method helps you to visualize where the things that are to be selected should go and how many other spaces (be they empty or for another object) should go. It seems like a trivial or inconsequential practice but ultimately, it helps, especially in the most complex of combinatorics problems (see Example 2.19, for example).

TIP 3 **Select the most restrictive object first.** If there is a strict restriction on a part of a combinatorial problem, it is logical to select that object first in order to ensure that it happens. Otherwise, unnecessary case work will cause you to spend too much time on one problem.

TIP 4 **Be methodical.** If you do need to work with cases in a combinatorical problem, make sure to work methodically. In problems with letters, work alphabetically. In problems with a number of things, start with 0 or 1 and go up one at a time from there. If you try case work without organization, you will often times lose sight of perhaps a crucial pattern that makes the whole problem very easy (again, see Example 2.19, for example).

TIP 5 **Counteract Over Counting.** Don't forget that often times, combinatorical problems have built-in instances of double counting or over counting. It's crucial that you see the double counting and subtract out those cases (see Example 2.18, for example).

9.3 Probability and Statistics Tips

TIP 6 **Note the difference between independent and dependent events.** In the probability question, decide which events are independent and which are dependent. In this way, you won't get confused about which to add and which to multiply.

TIP 7 **When working with averages (or means), work with totals.** Resist the urge to want to know *all* the data points. Often

9.4. SEQUENCES AND SERIES TIPS

times, it's enough just to know the total or sum of the data points. For example:

EXAMPLE 9.1 **All the students in an algebra class took a 100-point test. Five students scored 100, each student scored at least 60, and the mean score was 76. What is the smallest possible number of students in the class?**[1]

(A) 10   (B) 11   (C) 12   (D) 13   (E) 14

Solution: Let the number of students in the class be k. Therefore, the sum of the classes scores is $76k$. If the five scores of 100 are excluded, the sum of the remaining scores is $76k - 500$. And since each student scored at least a 60, the sum is at least $60(k-5)$. We can therefore set up the following inequality:

$$\begin{aligned} 76k - 500 &\geq 60(k-5) \\ 16k &\geq 200 \\ k &\geq 12.5 \end{aligned}$$

Since k must be an integer, the smallest k can be is $13 \implies$ (D).

9.4 Sequences and Series Tips

TIP 8 **Note carefully where a sequence starts and how many members it has.** Often times we look at a sequence and think we know where it starts, only to find that it starts on the next term or the term before. Similarly, we think we know it has n terms, when it really has $n - 1$ or $n + 1$ terms. Be careful when dealing with sequences.

TIP 9 **Any series worth summing infinitely is worth summing finitely.** G. K. Chesterton once said, "Anything worth doing is worth doing badly." Apply this quote to questions about series, that is, given an infinite series in a competition problem, chances are you don't really know to what it converges. However, if you start by summing the first few terms of the series, it might become clear, or clearer, where the series is going. Admittedly, you won't get the answer right every time, but you would be surprised how often this trick can win you points on a seemingly impossible series.

---

[1] from AMC12B 2004

9.5 Algebra Tips

TIP 10 **When working with exponential or logarithmic equations, try to standardize the bases.** It's difficult to see what's going on in an exponential or logarithmic equation if the bases are different. Therefore, use the laws of exponents and logarithms to change the bases so that the solution becomes clearer. Here's a simple example:

EXAMPLE 9.2 **Solve for x:**

$$2^{5x-3} = 8^x$$

Solution: The obvious base to standardize with would be 2. Change 8 into 2^3. Then

$$\begin{aligned} 2^{5x-3} &= \left(2^3\right)^x \\ 2^{5x-3} &= 2^{3x} \\ 5x - 3 &= 3x \\ x &= \frac{3}{2} \end{aligned}$$

9.6 Geometry Tips

TIP 11 **In circles where there is contact with other figures, always draw the radii to the points of contact.** When you draw the radii, you often see things like parallel lines, distances, or right triangles. The radii help you visualize the how the problem works and what trick should be used to solve it. In Example 6.1, for example, drawing the radii makes it apparent that the radius of the big circle is three times the radius of the small circles.

TIP 12 **In isosceles triangles, draw the altitude/median/angle bisector to the non-congruent side.** Like with radii in a circle, the altitude in the isosceles triangle helps you to visualize what's happening in the problem.

TIP 13 **Look for equilateral triangles, isosceles right triangles, 30–60–90 right triangles, and Pythagorean triples.** Often times, geometry problems will have these kinds of triangles hidden in the problem, and often times, they are the key to the problem. Know the 3 most popular Pythagorean triples: $3, 4, 5$; $5, 12, 13$; and $7, 24, 25$.

9.7 Trigonometry Tips

TIP 14 **KISS—Keep It Simple, Stupid.** Sometimes the complex trigonometric functions (tan, sec, csc and cot) mask what's going on in the problem. If you put everything in terms of sin and cos, the problem may become much easier to solve.

Chapter 10
Examples Cited

In this book, I took several examples from actual math competitions in addition to other mathematics textbooks. Here is a list of the resources used along with the online location in which they may be found.

- Marcus, Daniel A. *Combinatorics: A Problem Oriented Approach.* (Cambridge University Press, 1998).

- MATHCOUNTS Foundation, MATHCOUNTS School Competition, Sprint Round. Information and past competitions and solutions can be found online at

  http://www.mathcounts.org.

- Mathematical Association of America, AMC 12. Information about the MAA found online at http://www.mma.org. Problems and solutions found online at

  http://www.artofproblemsolving.com/Wiki/index.php/AMC_12_Problems_and_Solutions.

- Stanford University Mathematical Organization, Stanford Math Tournament. Information and past competitions and solutions can be found online at

  https://sumo.stanford.edu/smt.

- Texas A&M University, The Texas A&M University Math Contest. Information and past exams and solutions can be found online at

  http://www.math.tamu.edu/outreach/highschoolcontest.

- Whitworth, William Allen. *Choice and Chance with 1000 Exercises*. (Cambridge Deighton, Bell, 1901). Available online at the internet archive,

  https://archive.org/details/choicechancewith00whituoft

Index

$\sum a_i$, 49
φ, 55
$\varphi(n)$, 12
$_nR_k$, 22
30–60–90 right triangle, 70, 94

A-to-Z Formula, 50
acute triangle, 70
adjacent, 81
annulus, 77
arclength formulas, 66
area formulas
 circles, 66
 quadrilaterals, 68
 triangles, 70
arithmetic sequence, 78
average, 92

basic theorems
 circles, 66
Brahmagupta's Formula, 68

carrots, 25
casework (combinatorics), 31
central angle, 66
centroid, 72
Ceva's Theorem, 72
cevian, 72
change of base, 63
Choice and Chance, 22, 23, 41
combination, 20
 with repetition, 22
 without repetition, 21
complementary event, 37
conditional probability, 40

cone, 75
cone, right circular, 74
converge, 50
cosecant, 81
cosine, 81
 cosine of a sum, 82
 double angle formula, 83
 half angle formula, 83
cotangent, 81
cube, 75
cubic, 59
 maximum and minimum, 60
cyclic quadrilateral, 66, 68
cylinder, right, 74, 75

dashes and dividers, 22
Descartes' Rule of Signs, 60
diagonal formulas, 68
diagonal of a rectangular prism, 75
difference of cubes, 62
difference of squares, 61
difference of two consecutive squares, 54
Dirichlet, 27
discriminant, 57
distance to a line, 88
diverge, 50
divisibility, 7
division, inverted, 11
dodecahedron, 75
double angle formulas, 83
double counting, 28, 29, 32, 92
drawing altitude, 94
drawing radii, 94

$E(A)$, 39
equation of a circle, 90
equation of a parabola, 88
equilateral triangle, 70, 80, 84, 94
Euler's Formula for polyhedra, 76
Euler's Totient (or Phi) Formula, 12
event, 37
    dependent events, 92
    independent events, 40, 92
    probability of an event occurring, 37
    probability of dependent events, 41
    the ways an event may occur, 38
examples
    "words", 25
    association, 41
    Aaron and Moses, 42
    angles of a polygon, 78
    annulus, 77
    archer, 43
    arithmetic mean, 46
    arithmetic sequence, 78
    arithmetic series, 51
    averages, 93
    balls in bags, 33
    basketball teams, 21
    case work (combinatorics), 31, 33
    change of base, 64
    chewing gum, 30
    choice and chance, 42
    circles, 76, 77
    combinations with repetition/ "dashes and dividers", 23, 33
    combinations without repetition, 21
    cone, 79
    definition of logarithm, 63
    Descartes' Rule of Signs, 61
    divisibility rules, 8
    dominoes and dice, 43
    double counting, 29–31
    expectation value, 39
    flowers in a forest, 23
    from AMC12, 30, 58, 59, 62–64, 76–80, 84, 93
    from MATHCOUNTS, 46, 51
    from Stanford Math Tournament, 47
    from TAMU Math Contest, 52
    from *Choice and Chance*, 20, 23, 31, 33, 39, 41–43
    from *Combinatorics: A Problem Oriented Approach* by Daniel A. Marcus, 26
    GCD-LCM Product, 13
    grossing up, 52
    handkerchiefs, 23
    harmonic mean, 47
    hypotenuse as diameter, 78
    ice cream, 79
    inequality, 93
    infinite geometric series, 52
    law of cosines, 84
    letters, 20
    Mathuania, 19
    mean, 93
    Mississippi, 20
    mode, 44, 46
    movies, 15
    MPC, 18
    number of factors of n, 11
    organize, 41
    parabola, 59
    pentagon, 78
    permutation with duplicate objects, 20
    permutation with repetition, 20
    permutation without repetition, 19
    Pigeonhole Principle, 28
    prime factorization, 10, 64
    prime numbers, 51
    Principle of Inclusion-Exclusion (P. I. E.), 29, 30
    probability, 39

INDEX

proportional triangles, 78
quadratic, 58
quisquis, 31
recurrence relation, 26
relatively prime, 12
rhombus, 80
roots, 58
roulette, 39
running laps, 47
shirts and pants, 18
socks, 28
special polynomial factorizations, 62
special trigonometric values, 84
sphere, 79
spider, 30
stamps, 23, 26
sum of factors of n, 11
testing for primality, 9
totient, 12
trigonometric equations, 84
vertex, 59
expectation value, 39
exponents
    laws of exponents, 62
    standardization of bases, 94
exterior angle, 74

factoring formulas, 61
Fibonacci sequence, 26, 54
Fundamental Theorem of Algebra, 60
Fundamental Theorem of Arithmetic, 10

GCD-LCM product, 13
golden ratio, 55
greatest common divisor (GCD), 12

half angle formulas, 83
hangman method, 16, 92
Heron's Formula, 70
hypotenuse, 81

icosahedron, 75
index variable, 50

inequality, the geometric-arithmetic-harmonic, 46
inverted division, 11
isosceles right triangle, 70, 94

kite, 68

lattice points, 88
law of cosines, 83
law of sines, 83
Least Common Multiple (LCM), 13
logarithm, 63
    logarithmic formulas, 63
    standardization of bases, 94

matrix product, 90
mean, 44, 92
    arithmetic mean, 44
    geometric mean, 45
    harmonic mean, 45
median, 43, 72
    obtaining the median, 44
median concurrence theorem, 72
midpoint, 72
mode, 44
Multiplication Principle of Counting (MPC), 11, 16, 18, 21

n, 19, 21
$m \mid n$, 7
number of factors of n, 11
number of non-negative integer solutions, 24
number of positive integer solutions, 24

obtuse triangle, 70
octahedron, 75
odds (probability), 38
opposite, 81
opposite angles, 74

$p(A \cap B)$, 40
$p(A \cup B)$, 40
parallel lines, 74

parallelogram, 68
parcels, 24
partial sum, 49
percent change, 53
permutation, 16, 19
    with duplicate objects, 20
    with repetition, 19
    without repetition, 19
perpendicular diagonals, 68
perpendicular lines, 88
phi function, 12
Pick's Theorem, 88
Pigeonhole Principle, 16, 27
planar angles, 74
platonic solid, 75
point-slope form, 87
polygon
    sum of angles in a polygon, 74
power of a point, 67
prime, 9
    determining primality, 9
    prime factorization, 10, 91
    relatively prime, 11, 12
Principle of Inclusion-Exclusion (P. I. E.), 28, 29
probability, 38
Ptolemy's Theorem, 66
pyramid, 75
Pythagorean Theorem, 70
Pythagorean triple, 55, 94

quadratic
    maximum or minimum, 58
    quadratic formula, 57

range, 43
ranking, 19
rectangle, 68
rectangular prism, diagonal of, 75
recurrence relation, 26
reduced problem, 18
restrictive, 92
rhombus, 68
right triangle, 70, 81

roots
    product and sums of roots, 58, 59
rotation of a point, 90

secant, 81
selection, 20
sequence, 49
    arithmetic sequence, 50
    geometric sequence, 52
    number of members in a sequence, 93
series, 49
    arithmetic series, 50
    finite geometric series, 52
    infinite geometric series, 52
    summing the first few terms, 93
shoestring formula, 89
sigma notation, 49
Simple Addition Principle (SAP), 15, 18
sine, 81
    double angle formula, 83
    half angle formula, 83
    sine of a sum, 82
slope, 87
slope-intercept form, 87
SOHCAHTOA, 82
special features
    circles, 66
    triangles, 72
special sums
    sum of first n natural numbers, 53
    sum of the cubes of the first n natural numbers, 54
    sum of the first n odd numbers, 54
    sum of the squares of the first n natural numbers, 54
special trigonometric values, 82
sphere, 74, 75
square, 68
square of a sum, 61

INDEX

Stewart's Theorem, 72
subset, 25, 55
sum of cubes, 62
sum of factors of n, 11
surface area formulas, 74

tangent, 81
    tangent of a sum, 83
tangent (to a circle), 66, 67, 76
tetrahedron, 75
tiling, 55
totient, 12
trapezoid, 68
triangles, 70
two intercept form, 87

unity of probability, 37

vertex of a parabola, 88

Whitworth, W.A., 22–24
words, combinatorics, 25

years, the prime factorization of, 91

www.ingramcontent.com/pod-product-compliance
Lightning Source LLC
Chambersburg PA
CBHW071722170526
45165CB00005B/2112